Biotechnology Regulation and GMOs

BIOTECHNOLOGY REGULATION

Series Editor: Han Somsen, *Tilburg Institute of Law, Technology and Society (TILT), Tilburg University and Centre for Environmental Law, University of Amsterdam, The Netherlands*

Biotechnology is a term that provokes a range of differing reactions and views, though most would agree on the necessity of ensuring its careful regulation. The rise and rapid evolution of modern biotechnology has generated serious and diverse regulatory challenges, many of which have significant implications for society as a whole.

This series is designed to comprise work, both collaborative and single-authored, that provides a critical insight into the challenges presented by biotechnology, and the range of regulatory techniques and solutions on offer. The issues confronted in the series will range from agricultural biotechnology, including GMOs, to human genetics, through to intellectual property aspects of biotechnology, such as patents and the TRIPS framework. The European and international dimensions of biotechnology regulation will be a constant reference point.

Whilst focusing principally on the legal framework of biotechnology regulation, the series will also draw from related disciplines such as environmental studies, politics and biology, and aims to inform policy as much as to comment on it.

Titles in the series include:

Biotechnology Regulation and GMOs

Law, Technology and Public Contestations in Europe

Naveen Thayyil

Indian Institute of Technology, New Delhi, India

BIOTECHNOLOGY REGULATION

Edward Elgar

Cheltenham, UK • Northampton, MA, USA

Published by
Edward Elgar Publishing Limited
The Lypiatts
15 Lansdown Road
Cheltenham
Glos GL50 2JA
UK

Edward Elgar Publishing, Inc.
William Pratt House
9 Dewey Court
Northampton
Massachusetts 01060
USA

A catalogue record for this book
is available from the British Library

Library of Congress Control Number: 2013954338

This book is available electronically in the ElgarOnline.com Law Subject
Collection, E-ISBN 978 1 78347 388 5

ISBN 978 1 84844 564 2

Typeset by Columns Design XML Ltd, Reading
Printed and bound in Great Britain by T.J. International Ltd, Padstow

For my parents, Geeta and Kamal

Not just for all the sacrifices you made for us,
Also for teaching me through your life that it is Love,
and not materiality or gravity,
Which keeps this thing called the world together.

Contents

Series preface

This book is the third in a series that is devoted to the regulation of biotechnology. Since the series was launched, the EU has changed significantly, albeit in ways that are largely consistent with the vision laid out by the European Court of Justice in *Van Gend & Loos*. In that landmark case, the Court observed:

> Independently of the legislation of Member States, Community law not only imposes obligations on individuals but is also intended to confer upon them rights which become part of their legal heritage. These rights arise not only where they are expressly granted by the Treaty but also by reason of obligations which the Treaty imposes in a clearly defined way upon individuals as well as upon the Member States and upon the institutions of the Community.

The case marked a watershed moment for the EU, as it promotes individuals (in more contemporary parlance, 'European citizens') to the heart of the European project. Put differently, *Van Gend & Loos* signifies that the EU derives its legitimacy not from states or state bureaucracies but directly from its citizens. The future of Europe thereby is ultimately determined by individuals, who occupy a pivotal position in a web of duties *and* rights that they can exercise autonomously without state intermediaries. The emancipation of the European citizen continued after *Van Gend & Loos*, as is evidenced by Article 2 TEU:

> The Union is founded on the values of respect for human dignity, freedom, democracy, equality, the rule of law and respect for human rights, including the rights of persons belonging to minorities. These values are common to the Member States in a society in which pluralism, non-discrimination, tolerance, justice, solidarity and equality between women and men prevail.

It is striking that, despite its relatively recent origin, Article 2 does not reserve a discrete place for environmental values, but merely indirectly embraces these by way of anthropocentric references to 'human dignity' and 'respect for human rights'. This, in turn, brings us to Article 6(1) TEU, which recognizes the rights, freedoms and principles set out in the Charter of Fundamental Rights, and Article 6(3) which provides for fundamental rights, as guaranteed by the European Convention for the Protection of Human Rights and Fundamental Freedoms, and as they result from the constitutional traditions common to the Member States.

At a *fundamental* level, therefore, articulations of EU environmental policy that come in the form of directives and regulation thereby are manifestations of the right to a clean environment, which in Article 37 of the Charter is conceived as a solidarity right.[1] The anthropocentric paradigm underpinning EU environmental policy is important, as it bears upon the proper interpretation of secondary EU environmental law, in particular in the sphere of biotechnologies that affect the essence of the lives of humans and their relationships with their surroundings. The conceptualization in Article 37 of the Charter of the environment as a third generation human right implies that the EU and its Member States share duties to act, but also that citizens become the locus of both rights *and* duties. In his 1979 inaugural lecture, Karl Vasak put it as follows:

> [Third generation human rights] are new in the aspirations they express, are new from the point of view of human rights in that they seek to infuse the human dimension into areas where it has all too often been missing, having been left to the State, or States ... [T]hey are new in that they may both be *invoked against* the State and *demanded* of it; but above all (and herein lies their essential characteristic) *they can be realized only through the concerted efforts of all the actors on the social scene:* the individual, the State, public and private bodies and the international community.[2]

It is from this perspective that Naveen Thayyil's book is such a timely and important signal, confronting us with the challenge to uphold the values of Article 2 TEU in a high-tech globalized age driven by transnational technocracies that, in many ways, are prima facie incompatible with the respect for human dignity and pluralism Article 2 TEU professes. The author must therefore be applauded for this visionary book, in which he effortlessly combines legal, sociological and

---

[1] See Article 3(3) TEU, instructing the Union to pursue a high level of environmental protection, a mission the EU shares with the Member States (see Art. 4(2)(e) TFEU), and which is further specified in Articles 191–194 TFEU.

[2] Karel Vasak, 'Pour unetroisime generation des droits de l'homme', in *Studies and essays on international humanitarian law and Red Cross principles* 837, 839 (Christophe Swinarski ed., 1984).

philosophical arguments in support of a regulatory regime for genetically modified organisms that mirrors the fundamental values of the European Union.

Han Somsen
Tilburg, September 2013

Acknowledgements

This book is the product of my time at the Tilburg Institute of Law, Technology and Society, Tilburg University, the Netherlands. Gratitude is due to a number of people during this time at Tilburg, including my wonderful colleagues at TILT, right from Vivian Carter and Simone van der Hoff to Femke Abousalama and Geertrui van Overwallen. But it is only appropriate that I begin with the patient and witty Han Somsen, who fostered a sense of strategy in the preparation and writing of this book. Finishing this book would not have been possible without the kind and patient guidance of many others as well – Corien Prins, Bert van Roermond, Bert-Jaap Koops and Hans Lindahl all chipped in at various points during my fruitful years at Tilburg. Roger Brownsword at Kings College London, Maria Lee at University College London, Peter Fitzpatrick at Birkbeck, University of London and Sitaram Kakarala from the National Law School of India, Bangalore were all thankfully there at key points in the conceptualization and writing of this book. The book benefited considerably from the interactions I had in various reading groups, such a vital factor in the intellectual and academic vibrancy of any institutional space. A sense of warmth, solidarity and gratitude often accompany my recollections of the sessions (and colleagues in them), be it the Philosophy of Law seminar series at the Department of Philosophy in Tilburg, the Birkbeck fortnightly doctoral student reading sessions with Peter Fitzpatrick, the reading groups at TILT and the LASSNet (Law and Social Sciences Network) sessions at New Delhi and Pune.

Talking about solidarity, friends and other fellow travellers in Delhi and Bangalore fostered a sensibility of working towards a just and humane society; a sense that I hope has guided the writing of this book. Elizabeth VS, Prashant Bhushan, Usha Ramanathan, Babu Mathew, V Muralidhar, Anita Abraham and all other friends from Bhagwan Nagar; Govind Naidu, Siddharth Narrain and all other friends at the Alternate Law Forum, Bangalore, were all thankfully there at key points of my life over the last fifteen years, providing me with a sense of community, and helping me with their support and criticisms. Presence of dear friends and wonderful colleagues like Nadezdha Purtova, Esha Shah, Charlotte van Ooijen, Jas Groenedijke, Floor Fleurke and Viswa Reddy made the lonely path of a doctoral dissertation at Tilburg far less difficult than it could have been. Apart from these rewards of wonderful friendships, the encounter with Tilburg also blessed us with a beautiful name for our newborn, little Nadya. As important as anything else was the constant companionship and love of Shireen Mirza. We grew together with the writing of our respective doctoral theses – discussing, arguing, reading each other's drafts, criticizing and encouraging, and finding meaning in each other's lives; this itself makes me feel the journey of this book was worthwhile.

Last but not least, I am also very grateful for the insightful comments of the anonymous reviewers, as well as Victoria Litherland, Megan Ballantyne and Laura Mann at Elgar for their fantastic support and encouragement.

Abbreviations

AB	Appellate Body
ACP	African, Caribbean and Pacific Group of States
BBSRC	Biotechnology and Biosciences Research Council
BSE	Bovine Spongiform Encephalopathy
Bt.	*Bacillus thuringiensis*
CBD	Convention on Biodiversity
CJEU	Court of Justice of the European Union
DEFRA	Department for Environment, Food and Rural Affairs
DSB	Dispute Settlement Body
EC	European Community
ECHR	European Convention on Human Rights
ECJ	European Court of Justice
EFSA	European Food Safety Authority
EFTA	European Free Trade Association
EGE	European Group on Ethics in Science and New Technologies
EU	European Union
FAO	Food and Agriculture Organization
GAEIB	Group of Advisers on the Ethical Implications of Biotechnology
GATS	General Agreement on Trade in Services
GATT	General Agreement on Tariffs and Trade
GM	Genetically Modified
GMOs	Genetically Modified Organisms
ICTSD	International Centre for Trade and Sustainable Development
IMF	International Monetary Fund
IPR	Intellectual Property Rights
IVF	*In-vitro* Fertilization

LMO	Living Modified Organism
NBOs	National Bioethics Organizations
OECD	Organisation for Economic Co-operation and Development
OTA	Office of Technology Assessment
SPS	Sanitary and Phytosanitary
TAMI	Technology Assessment in Europe; Between Method and Impact
TBT	Technical Barriers to Trade
TEU	Treaty on European Union
TFEU	Treaty on the Functioning of the European Union
TRIPS	Agreement on Trade Related Aspects of Intellectual Property Rights
UK	United Kingdom
UNCTAD	United Nations Conference on Trade and Development
WB	World Bank
WHO	World Health Organization
WTO	World Trade Organization

Table of cases

EUROPEAN UNION

NATIONAL CASES

WTO: PANELS AND APPELLATE BODY

Table of legislation

Directives

Decisions

INTERNATIONAL INSTRUMENTS

1. Introduction

Not 'man' but 'men' inhabit the earth and form a world between them.[1]

This book examines how adequately European Union (EU) law treats serious contestations about the development and use of the radically new technology of genetically modified organisms (GMOs). This examination takes place against a backdrop of particularly strong misgivings about the regulatory decisions on GMOs, as well as intense disagreement about their necessity and desirability, which have paralysed their use in agriculture and food in the EU. In conventional frames of regulation of new technologies, decisions about the research, development and deployment of radical technologies like GMOs were to be left entirely to the scientific experts. Within this conventional frame, scientific experts are considered as best placed to understand, anticipate and decide on regulatory issues related to technologies like these. This book starts with an assumption that an approach that thinks of scientific experts as the starting point in such regulation is inadequate, for reasons that are elaborated later in this chapter. The widespread scepticism about underlying policy-making and results among the citizenry poses a serious strain to classical modes of representation, and necessitates a serious engagement with those outside the regulatory table.[2] One of the prominent ways in which this engagement with public contestations and representational issues has transpired is through policy commitments to make EU regulation more amenable to public participation. This book examines how adequately EU law has treated public participation as a means to represent and mediate public contestations about the use and regulation of GMOs.

Discussions about regulation of technologies like GMOs are often dominated by an obvious prescription, as for instance by Fukuyama, for the state to find the potential mix of benefits and threats inherent to

---

[1]   Hannah Arendt, *On Revolution* (Penguin 1963) 175. Emphasis added.

[2]   WRR Scientific Council for Government Policy, *Uncertain Safety: Allocating Responsibilities for Safety* (Amsterdam University Press 2008) 88. The representational aspect is raised 'also because of the formulation of alternative plans to regulate controversial technologies by concerned citizens'.

biotech development.[3] However, finding this middle ground is not a straightforward exercise since it is complicated by epistemic issues of knowability of both benefits and threats, regulatory capture by powerful interest groups or vocal agenda-setters, and the chasmic differences in cultural subjectivities and political normativities of groups who seek to shape regulation. Genetic modification in agriculture purports to increase yield, additive nutritive value or produce pesticide-like properties in crops through the technique of splicing – a process of deletion or insertion of identified genes for acquiring specific traits. Two decades of highly visible and controversial experience in EU regulation of genetically modified (GM) food and agriculture have provided important insights about conceptions of appropriate regulation of new technologies in general. GM crops have brought out fundamentally divided opinions about the appropriate development, use and regulation of the technology, throwing open profound philosophical and regulatory issues about modernity, rationality and progress. A plethora of literature in the past decade has focused on these differences through a variety of issues. They include the politics of production of seeds,[4] simultaneous increase in control and rent-seeking behaviour by large transnational corporations,[5] the domination of domestic regulatory structures by them, particularly in the Global South,[6] and seed sovereignty of marginal farmers in impoverished countries.[7] Other concerns have been raised during the regulation of GMOs, which are often considered technical, albeit with very significant political implications. They include concerns regarding the appropriate contours of employing the precautionary principle in controversial technologies;[8] the appropriate ambit of science in regulation of harm to

---

[3]   Francis Fukuyama, *Our Posthuman Future* (Picador 2002) 10.

[4]   Ralph Kloppenburg, *First the Seed: The Political Economy of Plant Biotechnology, 1491–2000* (University of Wisconsin Press 2004).

[5]   Jeremy Rifkin, *The Biotech Century* (Putnam 1998).

[6]   Mae-Wan Ho, *Genetic Engineering or Nightmare? The Brave New World of Bad Science and Big Business* (Gateway Books 1998).

[7]   Sheila Jasanoff, 'Biotechnology and empire: The global power of seeds and science' (2006) 21 *Osiris* 273.

[8]   Helena Valve and Jussi Kauppila, 'Enacting closure in the environmental control of GMOs' (2008) 20 *Journal of Environmental Law* 339. See further, Jim Dratwa, 'Taking risks with the precautionary principle: Food (and the environment) for thought at the European Commission' (2002) 4 *Journal of Environmental Policy and Planning* 197.

health and environment;[9] troubling doubts regarding toxicity, allergenicity and horizontal gene transfer as well as grave and irreversible effects on biodiversity from GMOs;[10] the increasing dominance of risk regimes in the governance of techno-scientific innovation;[11] the need for recognizing consumer voices in appropriate development of new technologies;[12] the issue of appropriate ambit of science and precaution in World Trade Organization (WTO) mechanisms;[13] and issues of culture and public opinion in regulatory regimes.[14]

Despite the visible domination of issues related to risk regulation in any brief survey of literature, it is pertinent to emphasize the breadth of the aforementioned issues.[15] The scope includes conflicting worldviews about danger and safety, understandings of ethical and environmental appropriateness in technological development and evaluation of social necessities in appraisals of technology, as well as the importance of political values in engaging with science and technology. Examining how regulators seek to navigate governance of the desirability of techno-scientific advancements is

---

[9] Bruno Latour, *The Politics of Nature: How to Bring the Sciences into Democracy* (Harvard University Press 2004).

[10] Artemis Dona and Ioannis S. Arvanitoyannis, 'Health risks of genetically modified foods' (2009) 49 *Critical Reviews in Food Science and Nutrition* 164; Christoph Then and Christof Potthof, 'Risk reloaded: Risk analysis of genetically engineered plants within the EU – A report by Testbiotech e.V. Institute for Independent Impact Assessment in Biotechnology' <http://www.testbiotech.org> last accessed 7 September 2011. See further, Jane Rissler and Margaret Mellon, *The Ecological Risks of Engineered Crops* (MIT Press 1996); Marc Lappe and Britt Bailey, *Against the Grain: The Genetic Transformation of Global Agriculture* (Earthscan 1999).

[11] Donatella Alessandrini, 'GMOs and the crisis of objectivity: Nature, science and the challenge of uncertainty' (2010) 19 *Social and Legal Studies* 3.

[12] Michel Callon, 'The increasing involvement of concerned groups in R&D policies: What lessons for public powers' in Aldo Geuna, Ammon Salter and Edward Steinmueller (eds), *Science and Innovation* (Edward Elgar 2003).

[13] Jacqueline Peel, *Science and Risk Regulation in International Law* (Cambridge University Press 2010).

[14] Sheila Jasanoff and Brian Wynne, 'Science and decision-making' in Steve Rayner and Elizabeth L. Malone (eds), *Human Choice and Climate Change* (Battelle Press 1998).

[15] See Maria Lee, *EU Regulation of GMOs: Law and Decision Making for a New Technology* (Edward Elgar 2008) 39–48. See for a general overview of the risk imperium, Gabe Mythen and Sandra Walklate, 'Introduction' in Gabe Mythen and Sandra Walklate (eds), *Thinking Beyond Risk Society* (Open University Press 2006) 3; Jane Franklin, 'Politics and risk', in Gabe Mythen and Sandra Walklate (eds), *Thinking Beyond Risk Society* (Open University Press 2006) 149.

related to the nature of the role that law plays in mediating and leading us through anxieties and hopes that accompany these conflicting world-views.[16] When techno-scientific communities and relevant regulatory structures are unable to acceptably mediate these public concerns, such cases require bringing rationalities from other sections of society to the regulatory table.[17] To make democratic decisions about which values should drive research, development and deployment of technology may, then, need further inputs rather than simply relying on scientific communities and the business establishments. It is in this context of recognizing the importance of democratic representation and values in making public decisions about technology, that the enlargement of arenas of regulatory deliberation is crucial. An important avenue for envisaging such enlargement of the arenas of deliberation has been through the commitment of public participation in EU policy documents. Whether such policy commitments are translated into justiciable legal principles and frameworks of public decision-making about GMOs needs further investigation.

The backdrop for this examination has significance over and above the straightforward concern of democratization of EU institutions and processes, by widening deliberative arenas of technology regulation through public participation. Linkages between law, technology and public contestations could have a crucial dimension in the shaping of democratic societies. A focus on contestations about technology, and regulatory choices between technology's various trajectories, is closely connected to the recognition that technological change is path dependent.[18] This recognition refuses to buy into the myths of techno-scientific change as an unravelling of a pre-ordained track of optimal progress and universal scientific reason. Such an approach brings focus on the spectres of technological slippery slopes that might not be appropriate for an

---

[16] Most commentators rightly rule out, at the outset, libertarian or laissez faire attitudes to technology development as misguided and unrealistic. That society should put constraints on the development of new technology to guide it in socially desirable ways is generally considered uncontroversial; see for a detailed exploration of this aspect: Janet A. Kourany, *Philosophy of Science after Feminism* (Oxford University Press 2010) chs 2 and 4.

[17] See for instance, Dahl's definition of democracy in terms of granting public contestation, where rationalities emanating from groups other than dominant governing groups can inform these contestations. Robert A. Dahl, *Polyarchy: Participation and Opposition* (Yale University Press 1971) 4.

[18] Path dependency assumes that technological development occurs along research tracks that are contingent to social and political choices, and are not naturally formed. See text to n. 60 for an elaboration.

ecologically sustainable economy, or might even be a debilitating tax on the possibilities of further democratization of our polities. Conceptually, notions of co-shaping of technology and society – that is, how social shaping of technology and technological shaping of society are inter-penetrative – underline the need to focus on treating contestations about technological trajectories as central to concerns of democratizing poli-ties.[19] The centrality of attempting social shaping of technology in a democratic manner, and for democratic ends, becomes an important starting point for enquiries about the normative role of law in the regulation of technology. Further, the technological trajectories that society chooses can shape social futures in significant manners. By way of substantiation, first, Jacques Ellul famously emphasized the need to seek ways of resisting and transcending technological determinism for preserving the very possibilities of freedom in the human condition. In his technological age, 'of autonomous, dominant totalizing systems of technology ... defined and dominated by it ... working as a principal law of our age',[20] he underlined the need to transcend the rationalities of techno-logical systems to protect human freedom. In other words, while one might want to avoid Luddite panic of rejecting all new technologies, how do we avoid human enslavement to technological rationality? Literary worlds from Shelley[21] to Atwood[22] focus on the dystopian possibilities of new and radical technologies like biotechnology leading to both authoritarian control and catastrophe, even when developed with good intentions, or even with hopes or assertions of great progress.

Competing political, economic and ethical constructions that respond to these concerns have played an important part in the kind of disagree-ments encountered in the EU regulation on GMOs. For instance, even for

---

[19]   See Wiebe Bijker and John Law (eds), *Shaping Technology/Building Society: Studies in Sociotechnical Change* (MIT Press 1992), in particular 'General introduction', 'Part I: Do technologies have trajectories?' and Thomas J. Misa, 'Controversy and closure in technological change: Constructing "steel"', 111. See further, Langdon Winner, 'Do artifacts have politics?' in Langdon Winner (ed.), *The Whale and the Reactor: A Search for Limits in an Age of High Technology* (University of Chicago Press 1986) 19.

[20]   Jacques Ellul, *The Technological Society* (John Wilkinson tr, Knopf 1964); *The Technological Bluff* (Geoffrey W. Bromiley tr, Eerdmans 1990). His was not a call for rejection of technique, but transcending it, by emphasizing our freedom through recognizing our non-freedom. See in particular, Ellul, 1964, xxxii.

[21]   Mary Shelley, *Frankenstein; or, The Modern Prometheus* (Harding, Mavor and Jones 1818).

[22]   Margaret Atwood, *Oryx and Crake* (Bloomsbury 2003); Margaret Atwood, *Year of the Flood* (Bloomsbury 2009).

many scientists the impasse in green biotechnology is a site to learn lessons about the appropriate development of technology, by taking societal concerns and ethical, legal and social aspects on board.[23] Thus an interest in how law takes into account public contestations of technology impacts not just the democratization of regulation, but also the development of appropriate technology in the future. In other words, this book's preoccupation relates to the conversational space constituted by law to articulate anxieties about our collective transition through technological pathways, and the accompanying contestation and disagreements about the regulation of these technological choices.[24] At this stage, it is pertinent to elaborate on the palpable nature of public concerns and unease about GMOs in the EU.

1.1 DISQUIET ABOUT GMOS: PERCEPTIBLE, CONSIDERABLE AND PERSISTENT

The introduction of GMOs in the EU as a product in the internal market has encountered high levels of public unease, disagreement and resistance, which is recognized by the multilevel regime in inordinate ways. In 1997, when the Commission authorized a variety of GM maize (on the back of a single positive vote from a Member State), there were angry

---

[23] Rip cites a number of statements from nanotechnology researchers to this effect. Arie Rip, 'The tension between fiction and precaution in nanotechnology' in Elizabeth Fisher, Judith S. Jones and Rene von Schomberg (eds), *Implementing the Precautionary Principle: Perspectives and Prospects* (Edward Elgar 2006) 270, 272.

[24] Here, I follow Brownsword's differentiation of law and regulation as intersecting but not coextensive. Law can be seen as a subset of regulation, as regulators rely on a number of instruments including law to influence behaviour of the regulated. At the same time regulation, which 'signifies something like: "the sustained and focused attempt to alter the behaviour of others according to standards or goals with the intention of producing a broadly identified outcome or outcomes, which may involve mechanisms of standard-setting, information-gathering and behaviour-modification," ... does not encompass such tasks as constitution making and dispute resolution'. Roger Brownsword, *Rights, Regulation and the Technological Revolution* (Oxford University Press 2008) 6–7. It is within this understanding that I assume law can set standards for appropriate regulation of technology, like, say, through setting legal requirements for public participation in such regulation.

reactions from a number of Member States;[25] and even a resolution by the European Parliament against it.[26] This resistance is continuing in a contingent manner visible in various ways that are elaborated here. A number of surveys and opinion polls, conducted by non-partisan and EU bodies, point towards substantial public resistance in accepting expert recommendations about the necessity, utility and safety of the technology. For instance, the *GM Nation* process initiated by the UK Government reported that only 8 percent of the respondents were happy with eating GM foods.[27] Importantly, public unease does not appear to be decreasing with the passage of time, contrary to a popular imagery of an ignorant public who can be enlightened with effective public education on relevant scientific and technological matters (what is derisively called PR techniques by some civil society groups).[28] An increase in information and education on the designed advantages from the development

---

[25]   See Tamara Harvey, 'Regulation of GM products in a multi-level system of governance: Science or citizens' (2001) 10 *Review of European Community and International Environmental Law* 321.

[26]   European Parliament, 'Resolution on genetically modified maize' final edition 08/04/1997.

[27]   See Department of Trade and Industry, *GM Nation? The Findings of the Public Debate* (UK Department of Trade and Industry 2003) para. 121, for more indicative details of the *GM Nation* process.

[28]   European Commission, *Attitudes of European Citizens Towards Environment* (Special Barometer 2952008) 64: 58 per cent of the respondents were found to be opposed to the use of GMOs. Earlier, in a 2005 survey, it was noted that 'while support may have increased between 1999 and 2002, it then decreased between 2002 and 2005', in George Gaskell, Nick Allum, Sally Stares and Agnes Allansdottir, *Europeans and Biotechnology in 2005: Patterns and Trends – Final Report on Eurobarometer 64.3* (European Commission 2006) 21. See further: Luc Bodiguel and Michael Cardwell, 'Genetically modified organisms and the public: Participation, preferences and protest' in Luc Bodiguel and Michael Cardwell (eds), *The Regulation of GMOs: Comparative Approaches* (Oxford University Press 2010) 11, 22; European Commission, *European Science and Technology* (Eurobarometer 55.22001); *European Science and Technology* (2nd edn, Eurobarometer 58.0 2003). Further, 'GMOs have been rejected as undesirable by the majority of European publics in successive polls in the last decade', Gregory Shaffer and Mark Pollack, 'Regulating between national fears and global disciplines: Agricultural biotechnology in the EU' (Jean Monnet Working Paper No. 10 2004) <http://centers.law.nyu.edu/jeanmonnet/papers/04/041001.pdf> last accessed 11 November 2011.

of GM technology does not seem to generate an increase in its popularity.[29] By 2010, around 169 regions and 4713 local governments in the EU had declared themselves GM-free.[30]

Public protest against the introduction of various GM crops has repeatedly taken the form of destruction of crops by protestors, notably in France, Belgium, the Netherlands and the United Kingdom. What is of interest here is not just the fact of destruction itself as an expression of public contestation, and accompanying issues of civil and political disobedience that pose a central problem to law's authority over citizens. What is of equal importance is the response of public juries in the UK to legal action against these destructions of property. These public juries have time and again acquitted such protestors. An example is the case of the prosecution of twenty-eight Greenpeace volunteers accused of crop destruction in GM fields, where the jury acquitted them of theft and failed to reach a verdict on criminal damage (19 April 2000). They were later acquitted regarding the latter charge (20 September 2000).[31] Similar is the instance of acquittal of volunteers charged with causing public nuisance by boarding a merchant vessel *MV Etoile*, which the volunteers claimed was carrying illegal animal feed comprising a mixture of non-GM and GM material from the USA.[32] Commentators have analysed this trend as an important indication of the public discomfort about criminalizing contestations, which most juries have generally considered as being in the public interest.[33]

---

[29]   This is clear in the report based on a series of polls commissioned by the Pew Initiative on Food and Biotechnology, a charitable trust established in 2001, with the aim of serving society as an honest broker in the debate on agricultural biotechnology, <http://www.pewtrusts.org/uploadedFiles/wwwpewtrustsorg/Reports/Food_and_Biotechnology/PIFB_StakeholderForum_Process.pdf> last accessed 29 October 2011.

[30]   GMO-free Europe, *GMO-Free Regions and Areas in Europe* (2009) <http://www.gmofreeregions.org/fileadmin/files/gmofreeregions/full_list/List_GMOfree_regions_Europe_update_September_2010.pdf> last accessed 1 December 2011.

[31]   See also *R. v Colchester* [2001] *Criminal Law Review* 564 as indicative of this trend of acquittals in the jury system, where it was the accused that were arguing for the possibility of greater damage, so as to secure the right to elect for jury trial; cited in Bodiguel and Cardwell (n. 28) fn 102.

[32]   British Broadcasting Corporation, *GM Protestors Cleared of Charges*, 16 September 2005; cf., Bodiguel and Cardwell (n. 28) fn 101.

[33]   'With juries being the representative of the *public*, this marks a perhaps unexpected engagement of wider society in the governance of GMOs', Bodiguel

Public concern is explicitly recognized in the recitals of the Aarhus Convention,[34] and recognition of public unease is crisply articulated by various legislatures in Europe, including the European Parliament.[35] The de-facto moratorium of the late 1990s that led to the breakdown of the previous EU regulatory regime was itself a product of the recognition by some national governments that public opinion in their territories expected them to oppose the inevitability of EU authorization of GMOs.[36] Under the current regime, Germany became in mid-2009 the sixth country to ban a particular strain of GM maize, despite prior clearance by EU institutions for crop cultivation all over Europe. Before this, Austria, Hungary, Poland, France and Greece had put in place safeguard bans on specific GMOs that had earlier been cleared by the requisite EU procedure.[37] These Member State actions hint at the possibility that the public concerns articulated in the de-facto moratorium

---

and Cardwell (n. 28) 26. See for a detailed comparison with the relevant French process, 26–35.

[34] UNECE Convention on access to information, public participation in decision-making and access to justice in environmental matters, 1998 <http://www.unece.org/fileadmin/DAM/env/pp/ documents/cep43e.pdf> last accessed 1 December 2011: 'Recognizing the concern of the public about the deliberate release of genetically modified organisms into the environment', Preamble of Aarhus Convention.

[35] European Parliament (n. 26).

[36] Two declarations were made in this regard in 2194 Council Meeting – 24/25 June 1999. *Declaration by the Danish, Greek, French, Italian and Luxembourg delegations concerning the suspension of new GMO authorisation; Declaration by the Austrian, Belgian, Finnish, German, Netherlands, Spanish and Swedish delegations.* There are various triggers identified for the moratorium; the most important include the strong and systematic picketing of supermarket outlets that sought to sell a new GM chocolate-bar brand in September 1998 that elicited 'no GM' policies from most European retail chains (that they would sell products only if there was a clear message from consumers that they really wanted these foods) and similar policies from large food manufacturers; the publication of a research report in *Nature* in May 1999, suggesting threat to biodiversity in data that pointed to adverse toxic effects of GM pollen to the larvae of the Monarch butterfly; as well as the media coverage and public reaction related to the immediate suspension and sacking of Arpad Pusztai, a scientist in the Rowett Research Institute in Aberdeen, UK, for issuing a press release regarding laboratory feed studies with rats that purportedly demonstrated increased toxicity in GM potatoes.

[37] European Commission, 'Communication from the Commission to the European Parliament, the Council, the economic and social committee and the Committee of the regions on the freedom for Member States to decide on the cultivation of genetically modified crops' COM (2010) 375 final, fn 4.

continue in the current framework.[38] The Council had rejected Commission proposals to repeal some of these bans by a qualified majority on four different occasions.[39]

Even the most trenchant critics of these bans within the EU, who characterize them as ill advised for a variety of reasons (including playing to populism, public ignorance and irrationality), have not ascribed mercantilism as a possible intent of these moves.[40] Significantly, there has been no such accusation of protectionism by Member States, including those who had originally voted for the release of the specific GMO. Some of the Member States which had originally voted for the release of individual GMOs have strongly resisted Commission proposals to revoke subsequent bans of other Member States. This is well illustrated by the EU voting on the Austrian withdrawal of a particular strand of GM maize,[41] where the EU President stated that most Member States had a second reasoning against the Commission's proposal, viz., 'the feeling that a Member State's will should be respected'.[42] Thus, though the reasons cited for safeguards included new scientific evidence of risk, or invoked inadequacies of risk appraisal to public concerns, the factum

---

[38]   '[I]t might be highlighted that lack of confidence in the regulatory framework extends well beyond the anti-GM movement, even to governments'; Bodiguel and Cardwell (n. 28) 36.

[39]   Commission Communication (n. 37) fn 6.

[40]   Indeed, allegations of trade protectionism from the other side of the Atlantic are commonplace, which is signified in the facts of the *EC-Biotech* dispute in the WTO. See text with n. 567 for a discussion.

[41]   Commission proposals for withdrawal of Austrian safeguard derogations were rejected twice by qualified majorities in the Environment Council, 2773rd Environmental Council Meeting, 18 December 2006, rejecting 'European Commission, Proposal for a Council Decision concerning the provisional prohibition of the use and sale in Austria of genetically modified maize', COM (2006) 510 final. In the vote on the second proposal the majority pointed out the failure to reassess the concerned GMOs under the Directive, and argued that 'the different agricultural structures and regional ecological characteristics in the EU need to be taken into account in a more systematic manner in the environmental risk assessment of GMOs'. The Commission's third proposal on the specific derogation only targeted the food and feed aspects of the safeguard action. However, even this proposal could not muster a support/rejection through qualified majority. This has left the Commission with the discretion to adopt its proposal – 2826 Environmental Council meeting, 14 November 2007.

[42]   <http://www.euractiv.com/en/biotech/eu-considers-pause-thought-gmos/article-168053> last accessed 13 June 2011.

of public opinion and political pressure to resist releases in specific Member States is intertwined with these invocations.[43]

Further, there appears to be a transition from the days of the biotech halcyon, when GMOs were hailed as a triumph of modern science, to a distinct recognition among decision-makers that public opinion and consumer voice are perhaps the most important aspect in the success of a technology like GM foods. Connected to consumer scepticism is the issue of a perceived lack of benefits for the general public; surveys show substantial sections of the public perceive the use of technology as mainly benefiting producers as opposed to ordinary people.[44] Literature surveys point out various points of convergence in the institutional pursuit of conversations with consumer voices. These include the finding that attitudes towards GM foods appear to be resistant to persuasion, 'including different arguments as absence of risks, improved risk assessment methods, introduction of traceability systems, public participation in risk management, and a whole range of producer and consumer related benefit; all to no avail'.[45]

The attempt to educate the public to allay their scepticism about specific technologies is based on a deficit model, which seeks to inform the ignorant lay person about the technical aspects and evaluation of risk. This model is founded on an assumption that lay people lack the cognitive ability to understand and appreciate issues of safety, and the basis of the public's irrational perception of risk – that is, the difference between a lay perception and an expert finding – is this information deficit. Within the deficit model, then, the problem of lay ignorance can be solved through effective communication of risk findings.[46] Deferring a

---

[43]   Commission Communication (n. 37) 6. Here the Commission remarked, in a matter of fact manner, that the Member State safeguard bans, though invoked as new scientific risks, in reality, involve diverse and varied types of reasons, including political or economic motivations such as meeting the demand for GM-free markets.

[44]   *GM Nation* (n. 27) para. 121. Earlier, the UK Food Safety Authority survey in 2000 found that the presence of GM ingredients was a key criterion for consumers in their decision to avoid purchase from takeaways: House of Lords Select Committee, 2000, fns 4–5, cited in Michael Cardwell, 'The release of genetically modified organisms into the environment: Public concerns and regulatory responses' (2002) 4 *Environmental Law Review* 156.

[45]   Joachim Scholderer, 'The GM foods debate in Europe: History, regulatory solutions, and consumer response research' (2005) 5 *Journal of Public Affairs* 263, 270.

[46]   Susan Owens, 'Engaging the public: Information and deliberation in environmental policy' (2000) 32 *Environment and Planning* 1141, 1142.

discussion on the deficit model of risk to the third chapter, here I only point to a major criticism regarding the assumption that responses and judgements about dangers are purely a mechanical processing of information. Instead, this cognitive view of a human being is criticized as too much of a simplification. It has been asserted that people forge questions and find answers about issues that concern them, rather than purely perceive and process the information that is provided to them.[47] Various scholars have underlined the importance of 'affect heuristics' when people make judgements.[48] Slovic noted: 'Although risk perception was originally viewed as a form of deliberative, analytic information processing, over time we have come to recognize how highly dependent it is upon intuitive and experiential thinking, guided by emotional and affective processes'.[49] It appears that there is a gradual recognition that public beliefs and consumer evaluations are highly complex, and cannot be merely encapsulated by ascriptions of safety and risk, though safety heuristics are often used to ascribe rationality to visible public unease. Issues of general socio-political attitudes (including those towards the environment and nature), increase of corporate control over science, technocratic regulation and the resultant lack of social trust were all found to be connected to public articulations of safety and risk in agricultural biotechnology.[50] This state of public opinion is understood to arise partly from the large-scale appearance of issues related to food safety and bioethics in public agendas, including through considerable efforts of green and consumer groups. The Bovine Spongiform Encephalopathy (BSE) crisis and media coverage of how regulators handled it, as also the anxieties from the sudden appearance of Dolly the cloned sheep, could only have reinforced this public scepticism.

For reasons mentioned in this section, it is clear that public unease about GMOs is clear and considerable, and the contestations about opting for this technological trajectory perceptible. Further, persistent opposition is despite regulatory attempts at science communication and consumer education, attempts that are based on a flawed understanding that disagreement solely reflects the ignorance of the public. The continuation

[47]   Serge Moscovici, 'The phenomenon of social representations' in Robert Farr and Serge Moscovici (eds), *Social Representations* (Cambridge University Press 1984) 3.

[48]   See for an elaboration: Helene Joffe, 'Risk: From perception to social representation' (2003) 42 *British Journal of Social Psychology* 55, 58.

[49]   Paul Slovic (ed.), *Risk Perception* (Earthscan 2000) xxxi.

[50]   Michael Siegrist, 'The influence of trust and perceptions of risk and benefits on the acceptance of gene technology' (2000) 20 *Risk Analysis* 195.

and accentuation of public contestations to the use of GMOs highlight the inadequacies with representation in the regulation of radical technologies. This is since democratic law ought to frame principles to reliably ensure that policies are responsive to citizens' preferences, albeit within the legitimate ambit of constitutional values. The normative and consequentialist reasons for being responsive to citizens' preferences, including through public participation, are elaborated in the next section.

1.2 BENEFITS FROM PUBLIC ENGAGEMENT WITH CONTESTATIONS ABOUT TECHNOLOGY

Should law require regulation of new technologies to be as proximate and responsive to citizens' preferences as possible is an important question that needs consideration. A serious engagement with public contestations about technologies, and their trajectories, is beneficial for the democratization of both technology and law. To start with many prominent sociologists would insist that the design of any technology, as it is finally deployed in the market, cannot be understood without the political, professional and economic factors that have shaped it at the drawing board; 'technologies are born out of conflict, difference or resistance'.[51] Resistance to particular kinds/trajectories of technology is identified as having an important signalling function in the development of technology:

> Rather than asking what causes people's reactions to new technology, we ask how their reactions influence the development of new technology. Resistance is an independent variable of the design rather than a dependent one ... resistance is not about changing or not changing, but about the direction and the rate of change ... resistance is more a problem of the designer than of the user. But in avoiding the trap of shifting the attribution from user to designer, one needs to see the systemic process between designers and users over time... where designers are not the only actors, there is capital involved ... shareholders, buyers and sales-people, all following a different logic. Highlighting the contributions of resistance to that 'game' empirically in different contexts serves to enhance the user perspective.[52]

---

[51] Bijker and Law (n. 19) 9.
[52] Martin Bauer, 'Technophobia: A misleading conception of resistance to new technology' in Martin Bauer (ed.), *Resistance to Technology* (Cambridge University Press 1995) 97.

User perspectives and the way technology is developed can have various ramifications beyond the perspectives of an individual consumer, as larger questions of governance and citizenship are intricately connected to such signalling during development of technology.

There is a pejorative trend of ascribing contestations and resistance to particular technologies, like GMOs, as technophobic and irrational. In contrast, the EU Expert Group on Science and Governance points out the possibilities of existence of alternate rationalities of public unease hidden from dominant scientific paradigms:

> [P]erhaps the most widely recognized indicator of public unease concerns reactions to issues at the intersection of 'science' (including science-based technologies) and 'risk'. The public is thought to fear science because scientific innovations entail risk. Both science and risk, however, are ambiguous objects. It is frequently assumed in policy circles that the meanings of both for citizens must be the same as for experts, but that assumption is, in our view, itself a key element in generating 'public unease'. The widespread sense of unease – sometimes expressed as 'mistrust of' or 'alienation from' science – must be seen in broader perspective. We conclude indeed that there is no general, indiscriminate public disaffection with or fear of 'science'. Instead, there is selective disaffection in particular fields of science, amidst wider areas of acceptance – even enthusiasm.[53]

Further, it could be that:

> [a]lthough the European public has often been characterized as technophobic, emotional or susceptible to NGO and media propaganda over GMOs, there is nothing irrational about disagreeing with a scientific estimate of risk. People may not know the technical details, but they have developed awareness of the broad issues involved and ways of forming their own judgments. Their previous experience, such as the BSE crisis, found concerns regarding the (un)reliability of scientific reassurances of safety and the authorities' tendency to conceal information when powerful economic interests are at stake.[54]

At one level, the significance for regulators to take into account overt contestations about technologies like GMOs could be to ensure the most

---

[53]   Expert Group on Science and Governance, *Taking European Knowledge Society Seriously: Report of the Expert Group on Science and Governance to the European Commission* (EUR 22700 2007) 9 <http://ec.europa.eu/research/sciencesociety/document_library/pdf_06/europeanknowledge-society_en.pdf> last accessed 21 December 2011.

[54]   Elsa Tsioumani, 'Genetically modified organisms in the EU: Public attitudes and regulatory developments' (2004) 3 *Review of European Community and International Environmental Law* 279, 281.

responsible designs of technologies are developed and adopted in society. Further, there could be reasons related to pursuing democratic ideals that necessitate regulators to take into account such voices. Important scholarship on scientific citizenship and governance of new technologies conceptually connects general concerns about disagreements regarding law and justice to contestations about technological trajectories. In this stream, science, politics and law are seen to be closely entangled, and regulation of science and technology has been identified as a major challenge for democratic governance, 'since political decisions about science and technology inevitably entail questions of democracy'.[55] First, the way certain technologies are developed or deployed may lead to more authoritarian states or can dramatically concentrate wealth and power among smaller sections of a society. For instance, the physical arrangements that modern technology effectuated through industrial production, war and communication have fundamentally changed exercise of power and experience of citizenship.[56]

Secondly, the way in which law and regulation structures/makes decisions about specific technologies can have general effects on the polity in a manner that is over and above how such technologies change society. How regulators treat contestations about technology, especially in cases of fundamental disagreements and persistent contestations, can have important effects on the law's aspiration to be democratic. Engaging with public unease about specific technologies could possibly be approached by some regulators to merely engineer legitimacy, where the expression of public unease becomes only a practical hurdle that needs to be overcome. However, such an approach to regulation of technology belies law's general claim to be genuinely democratic. Beyond proceduralities of decision-making, democratic normativity requires law to have a serious engagement with the inescapability of disagreement about even fundamental matters of law, and not a caustic acknowledgement about

---

[55] See for instance: Alfons Bora, 'Techno-scientific normativity and the "iron cage" of law' (2010) 35 *Science, Technology and Human Values* 3. He examines a plethora of literature to make this observation including Yaron Ezrahi, *Descent of Icarus: Science and the Transformation of Contemporary Society* (Harvard University Press 1990); Steve Fuller, *The Governance of Science: Ideology and the Future of the Open Society* (Open University Press 2000); Daniel Lee Kleinman (ed.), *Science, Technology and Democracy* (State University of New York Press 2000); Philip Kitcher, *Science, Truth and Democracy* (Oxford University Press 2001); and Sheila Jasanoff, *Science at Bar: Law, Science and Technology in America* (Harvard University Press 1995).

[56] Winner (n. 19) 20.

disagreement in views about religion, ethics and philosophy.[57] Waldron points out that 'our common basis for action in matters of justice has to be forged in the heat of our disagreements, not predicated on the assumption of a cool consensus that exists only as an ideal'.[58] He distinguishes this from a general assumption and argues that such an inaccurate common view cannot obliterate the fact of disagreement.[59] Given the fact that widespread contestations about regulation of GMOs have persisted for decades in the EU, perhaps based on vantage-points that are different from dominant regulatory and scientific communities, it becomes important for regulators to treat these disagreements seriously so as to make a claim for democratic normativity. It would be difficult to deny the reasonableness of a normative expectation on law to take disagreements seriously, especially in situations where substantial sections of citizens (perhaps even a majority) consistently appear to disagree with the release of GMOs. Hence it would be reasonable to start from a position that, once there is an assumption of considerable and persistent disagreement, the representational question in regulatory deliberation needs to be taken more seriously, including through ensuring effective participation of publics which are wider than expert communities. It could be missing a trick to ignore public unease with a hope that the public will turn around, or by explaining unease away as ascriptions of irrationality of publics, or even as mere products of agenda-setting by civil society organizations.

A serious engagement with societal disagreement about technologies is also directly relevant to the development of democratic technology. The assumption here is to move away from an understanding of technological change as a race along a single pre-ordained track following a single self-evidently optimal line of progress, to one where technological

---

[57]   Jeremy Waldron, *The Dignity of Legislation* (Cambridge University Press 1999) 154–155.

[58]   During a sympathetic account of political process in Western liberal democracies, Waldron recounted that 'every single step that has been taken by legislatures towards making society safer, more civilized, and more just has been taken against a background of disagreement, but taken nevertheless in a way that managed to somehow retain the loyalty and compliance (albeit often grudging …) of those who in good faith opposed the measures in question … including prohibition of child labor, limitation of working hours, dismantling of segregation, health and safety regulation in factories, liberation of women', as opposed to a well-ordered Rawlsian justice consensus. He emphasized that these political achievements had authority and respect as law, despite disagreements even as to whether it is a step in the right direction, ibid 156.

[59]   Ibid.

development is more like biological evolution. Development of technologies is better understood as an 'open branching process more akin to organic growth', and interacting with 'wider social structures and contextual contingencies to become channelled in highly path-dependent ways'.[60] Recognition of path dependency, and a refusal to buy into the myths of technological change as an unravelling of a pre-ordained track of optimal progress, make a focus on law's engagement with disagreement an important way to identify and deal with spectres of technological lock-in. Or, as Charles Taylor would suggest, 'we are not indeed locked in. But there is a slope, an incline in things that is all too easy to slide down'.[61] Once we recognize that the development of technologies and their trajectories is not linear and natural, but is contingent on social factors, the influence of user resistance in social shaping of technology is

---

[60] See for a succinct introduction to this position: Andy Stirling, 'Science, precaution and the politics of technological risk: Converging implications in evolutionary and social scientific perspectives' (2008) 1128 *Annals of the New York Academy of Sciences* 95, 97. He continues: 'In a complex, dynamic, interconnected, and finite world, only a small subset of the totality of potentially viable developmental pathways will actually be followed. This can be as true at the level of the design of an individual consumer product, like the video or DVD, as at the global scale taken in the configuration of major infrastructures, like those underlying energy, transport, communication, and industrial production systems'. Questions of path dependency are often dealt away for various reasons, mainly, I suspect, due to the dilemma of regulatory connection, where we know very little about the technology at the initial stages of its development, and at a later stage, when we get to know about, it is too late to stop it; see Paul Sollie, 'Ethics, technology development and uncertainty: An outline for any future ethics of technology' (2007) 5 *Journal of Information, Communication and Ethics in Society* 293, 297.
Falsities and inadequacies of the assumption behind a single and self-evident optimal line of progress unravelling technological developments are often ignored through various heuristic mechanisms including: avoidance – for instance, Calestous Juma, 'The new culture of innovation: Africa in the age of technological opportunities', Keynote Address (8th Summit of the African Union, Addis Ababa, 29 January 2007); wishing it away – for instance, Mireille Hildebrandt, 'A vision of ambient law' in Roger Brownsword and Karen Yeung (eds), *Regulating Technologies: Legal Futures, Regulatory Frames and Technological Fixes* (Hart Publishing 2008) 175, 176; as also general assertions that it is indeed such unravelling of the universal scientific rationale that unfolds through technological development – for instance, Paul Ceruzzi, 'Moore's Law and technological determinism: Reflections on the history of technology' (2005) 46 *Technology and Culture* 584, 593.
[61] Quoted in Richard Spinello, *Cyberethics: Morality and Law in Cyberspace* (3rd edn, Jones and Bartlett 2006) 9.

evident.[62] Such social shaping includes mediation of the nature and effects of slippery slopes that have to be identified early enough to avoid situations of a fait accompli. For instance, fear of irreversibility is visible in the claims of some public groups that large scale cultivation of GM crops leads to lock-in of agriculture to GM crops due to what they see as unmanageable mixing of genetic material between GM and other kinds of agriculture. They quote the experience of widespread cultivation of soya in Argentina or canola in Canada for the slim possibilities of returning to an organic or GM-free form of cultivation, once widespread transgenic cultivation is allowed.[63] These pictures of irreversibility and lock-ins are differently mediated by different regulatory communities, where the representation and participation of farming and other public groups are important. Whether law specifies scope for participation of these groups, which are traditionally outside the pale of regulatory deliberation, impacts the evaluation and management of ecological slippery slopes.

1.3 DEMOCRATIC DEFICIT IN POSTNATIONAL FRAMEWORKS

The general issues that make the legal engagement with public contestations about GMOs important appear to be accentuated in postnational frameworks like the EU.[64] Concerns of democratic deficits in the EU

---

[62] Compelling evidences of how social change and technological change shape each other (co-construction of technology and society) are considered by many philosophers of technology. See Wiebe Bijker, *Of Bicycles, Bakelite and Bulbs: Toward a Theory of Sociotechnical Change* (MIT Press 1995); Wiebe Bijker and John Law, 'Do technologies have trajectories?' in Bijker and Law (n. 19) 15; Thomas J. Misa, 'Controversy and closure in technological change: Constructing "steel"' in Bijker and Law (n. 19) 109.

[63] Gundula Meziani and Hugh Warwick, *Seeds of Doubt* (Soil Association 2002). See further, Kahon Chan, "War of the papayas", Do technologies have trajectories?' *China Daily* (Hong Kong, 8 September 2011) <http://www. chinadaily.com.cn/hkedition/201109/08/content_13645581.htm> last accessed 21 September 2011.

[64] The term 'postnational' is used interchangeably with 'supra-national' here, and means legal spaces that have emerged as a result of the decline of the dichotomy between national and international law, triggered by increasing inter-linkages in structures of national/regional and global governance. See further: Nico Krisch, *Beyond Constitutionalism: The Pluralist Structure of Post-national Law* (Oxford University Press 2010).

institutional architecture are persistently raised without resolution, not-withstanding robust defences of the EU regulatory regimes by renowned political scientists like Majone and Moravcsik.[65] These concerns include issues such as how direct an influence are EU citizens' policy preferences on EU policy outcomes, or, in other words, how technocratic or political is the policy process.[66] Here, an important query concerns the existence of opportunities in the institutional design that allow for an opposition to the current leadership elites and policy status quos, an essential feature in liberal democracies.[67] There have been various institutional attempts to address the important question of the democratic deficit, including recent attempts at reform of delegation and implementation powers of the Commission.[68] Whether these attempts can resolve such concerns is yet to be seen.[69] Nevertheless, an architecture that allows the release of GM seeds despite a majority of Member States repeatedly voting against it in the committee procedure,[70] as well as substantial contestation and unease among the general population,[71] accentuates already existing concerns of

---

[65]   Francesca Bignami, 'The democratic deficit in the European Community rulemaking: A call for notice and comment in comitology' (1999) 40 *Harvard International Law Journal* 451. See for Majone's regulatory state thesis: Gian-domenico Majone (ed.), *Regulating Europe* (Routledge 1996) and Giandomenico Majone, 'Europe's democratic deficit' (1998) 4 *European Law Journal* 5. See for Moravcsik's inter-governmentality thesis: Andrew Moravcsik, 'Despotism in Brussels? Misreading the European Union' [2001] *Foreign Affairs* 114.

[66]   See Andreas Follesdal and Simon Hix, 'Why there is a democratic deficit in the EU: A response to Majone and Moravcsik' (2006) 44 *Journal for Common Market Studies* 533, 534.

[67]   Ibid 548.

[68]   See the co-decision procedure by Council and the European Parliament for legislative acts under Article 289 of the Treaty on the Functioning of the European Union (TFEU), procedures for delegated acts (Article 290 TFEU) and procedures for implementing acts (Article 291 TFEU). Further, see Regulation 182/2011 of the European Parliament and of the Council laying down the rules and general principles concerning mechanisms for control by Member States of the Commission's exercise of implementing powers [2011] OJ L55/13.

[69]   Vihar Georgiev, 'Commission on the loose? Delegated lawmaking and comitology after Lisbon' (EUSA Twelfth Biennial International Conference, Boston, March 2011); Andreas Follesdal, 'The legitimacy challenges for new modes of governance: Trustworthy responsiveness' (2011) 46 *Government and Opposition* 81; Christopher Lord, 'Still in democratic deficit' (2008) 43 *Inter-economics: Review of European Economic Policy* 316.

[70]   Text near n. 162.

[71]   Text near nn. 25–45.

democratic deficit. Public participation has been offered as an institutional route to address this concern of representational and democratic deficit (as discussed later in this chapter, as well as in the second chapter). By focusing on the room for public participation in EU regulation of GMOs, this book seeks to examine how adequately EU law treats serious disagreements about the development and use of radical technologies like GMOs.

1.4   CHARACTER OF SCIENTIFIC ADVICE IN THE REGULATION OF TECHNOLOGY

Issues of public participation, representation and democratization in the regulation of technologies like GMOs are intricately connected to how law involves scientific advice. Since scientific advice plays an important part in the regulation of technology, especially in risk regulation, it requires us to bracket an appropriate understanding of the contemporary scientific enterprise. The broad sociological literature that has characterized contemporary scientific enterprise as *Mode-2*, *post-normal* or *post-academic*, as opposed to normal or traditional science, is of importance here.[72] These enumerated approaches understand science as beyond its traditional understanding, of say, Kuhn.[73]

Kuhn famously characterized scientific enterprise (*normal* science) as a process regarding debates about the rules of science, 'research firmly based upon one or more past scientific achievements, achievements that some particular scientific community acknowledges for a time as supplying the foundation for its further practice'.[74] He saw scientific revolutions as 'tradition shattering complements to the tradition-bound activity of normal science', which predicates that even scientific change is bound by the internal heuristics of a discipline.[75] As opposed to this *normal* condition, contemporary scientific research and expertise is marked by a number of factors mentioned below, which have implications for the use of scientific expert advice in the regulation of technology. Where 'facts

---

[72]   Though there are internal differences among these approaches that characterize science, all of them have similar effects for the limited purpose here – which is related to employment of scientific expertise in regulation, enumerated in the subsequent paragraphs.

[73]   Thomas Kuhn, *The Structure of Scientific Revolutions* (University of Chicago Press 1962).

[74]   Ibid 10.

[75]   Ibid 6.

are uncertain, values in dispute, stakes high and decisions urgent' the conventional model of the linear communication of the findings of scientific enquiry to policy-makers may be insufficient.[76]

> [N]ormal science aims at establishing the ultimate truth or the final resolution of a scientific puzzle, post-normal science recognizes that as long as both scientific uncertainties and decision-stakes are high, such aim is in principle unachievable. Indeed it can be misleading and create false expectations to act as if the role of science in such issues is just to get the facts right. Instead post-normal science aims at common commitments to reflective approaches for dealing with complex policy issues.[77]

Various sociologists of science have asserted the differences between the nature of the contemporary scientific endeavour and the earlier ideal-type of science. The classical ideal-type uses experimental systems that consist of only one epistemic object (or unknown element), while the former research is faced with a multi-sited heuristic endeavour having multiple epistemic objects. This makes contemporary scientific research, which informs regulation, a multi- or interdisciplinary confrontation. This negates general expectations of objective standards and absolute truths during risk regulation. Another difference identified by these scholars is that while in the first case (*normal* science) the disciplines are primarily scientifically framed, in the second case the individual scientific fields also fulfil a societal function and follow explicit trans-scientific object-ives, often leading these multiple trans-scientific objectives to be in conflict with each other.[78] Here, then, the nature of advice that emanates from scientific enterprises depends on the character of the trans-scientific objective.

Parallel is the characterization of contemporary science as Mode-2, which notes the replacement of traditional university-oriented forms of

---

[76] Jerome Ravetz, 'Usable knowledge, usable ignorance: Incomplete science with policy implications' in William C. Clark and Robert Munn (eds), *Sustainable Development of the Biosphere* (Cambridge University Press 1986) 415, 422. See further Silvio Funtowicz and Jerome Ravetz, 'Science for the post-normal age' (1993) 25 *Futures* 735.

[77] Arthur Petersen, Albert Cath, Maria Hage, Eva Kunseler and Jeroen van der Sluijs, 'Post-normal Science in practice at the Netherlands Environmental Assessment Agency' (2011) 36 *Science, Technology and Human Values* 362, 367.

[78] Silvio Funtowicz and Jerome Ravetz, 'Post-normal science: An insight now maturing' (1999) 31 *Futures* 641.

mono- or interdisciplinarity by *trans-disciplinary* science that 'is embedded in non-hierarchical, heterogeneous, transient settings'.[79] The blurring of scientific disciplines is seen to be accompanied by a commonplace mixing of what is traditionally understood as separate enterprises of scientific, technological and industrial research, both outside and inside universities.[80] In this scenario, which is also termed 'post-academic', the characteristics of disinterestedness and objectivity of the scientific enterprise cannot be taken for granted any more.[81] The large-scale industrialization of science for industrial, economic and social application, and the accompanying involvement of concerns of profit in corporate investment, are seen to affect research agendas of the scientific enterprise. The Dutch scientific council for government policy (WRR) refers to the fading distinctions between doing science and doing business as implicating the position of expert advice in regulation.[82] These characterizations of contemporary scientific expertise, and of its use in regulation, are widely recognized as valid by sociologists of science.[83] Therefore, for an adequate understanding of scientific advice in regulation to deal with the representational problem, one may additionally need the consideration of public values; an aspect that is elaborated in Chapters 3 and 6. Here public participation is often claimed as an important avenue for such ascertainment of appropriate values in regulation by EU bodies.

---

[79] Helga Nowotny, Peter Scott and Michael Gibbons, *Rethinking Science: Knowledge and the Public in an Age of Uncertainty* (Polity Press 2001); Michael Gibbons, *The New Production of Knowledge: The Dynamics of Science and Research in Contemporary Societies* (Sage 1994).

[80] Robert Hagendijk, 'The public understanding of science and public participation in regulated worlds' (2004) 42 *Minerva* 41.

[81] John Ziman, 'Is science losing its objectivity?' (1996) 382 *Nature* 751; John Ziman, *Real Science: What It Is, and What It Means* (Cambridge University Press 2000).

[82] WRR (n. 2) 85–87.

[83] See a broad review of literature regarding post-normal science as focus of critique in John Turnpenny, Mavis Jones and Irene Lorenzoni, 'Where now for post-normal science? A critical review of its development, definitions, and uses' (2011) 36 *Science, Technology and Human Values* 287, 295. See further, Sheila Jasanoff, 'Breaking the waves in Science Studies: Comment on H.M. Collins and Robert Evans, The Third Wave of Science Studies' (2003) 33 *Social Studies of Science* 389.

1.5 EU INVOCATION OF PUBLIC PARTICIPATION

Arguments in favour of public participation include the cognitive improvement of decisions due to inputs from a plurality of perspectives, including widening its technical and scientific base.[84] Further, they also include an implied emphasis on participation as furthering justice and equity,[85] ambitions to make participative or deliberative measures supplements or alternatives to representative democracy,[86] enhancement of legitimacy of controversial environmental decisions that are frequently delegated to unelected experts,[87] and normative advantages for representative democracies through improvements in transparency, accountability and implementation, resulting in a wider range of democratic deliberation.[88] While acknowledging possible contradictions between these multiplicities of aspirations, the reason for employing public participation as a marker in this book is discussed in this section.

The acceptance of public participation as a necessity in important EU policy documents lends impetus for an investigation about how law provides for public participation in EU GMO regulation. In 2001, the Commission's White Paper on European Governance signified the need to make policy-making more inclusive through stronger interaction with civil society.[89] It recommended 'wide participation throughout the policy chain – from conception to implementation' to ensure the 'quality, relevance and effectiveness of EU policies'.[90] A similarly strong invocation of people-centred policies is found in the Life Sciences strategy of 2002, where the Commission set out its vision of developing 'sustainable and responsible policies' in healthcare, agriculture and food production,

---

[84] Jenny Steele, 'Participation and deliberation in environmental law: Exploring a problem-solving approach' (2001) 21 *Oxford Journal of Legal Studies* 415.

[85] See for instance, World Commission on Environment and Development, *Our Common Future* (Oxford University Press 1987).

[86] Maria Lee and Carolyn Abbot, 'The usual suspects? Public participation under the Aarhus Convention' (2003) 66 *Modern Law Review* 80, 83; Simon Chambers, 'Deliberative democratic theory' (2003) 6 *Annual Review of Political Science* 307, 308.

[87] Ibid, Lee and Abbot 84.

[88] Jon Elster (ed.), *Deliberative Democracy* (Cambridge University Press 1998); John Dryzek, *Deliberative Democracy and Beyond: Liberals, Critics, Contestations* (Oxford University Press 2002); John Dryzek and Simon Niemeyer, *Foundations and Frontiers of Deliberative Governance* (Oxford University Press 2010).

[89] COM (2001) 428 final, 25 July 2001.

[90] Ibid 10.

including in agricultural biotechnologies.[91] However, both documents deny the necessity of translating this expression of interest in public participation into justiciable principles.[92] Notwithstanding this, the acceptance of the necessity of public participation in EU governance in these documents is important for our purpose.[93]

A number of international legal instruments underline the importance of public participation, the need to institutionalize a regulatory framework that allows effective participation in public issues, and even recognize its special relevance in the case of GMOs. Principle 10 of the Rio Declaration calls for the provision of appropriate access to information, the opportunity to participate in decision-making processes, and effective access to judicial and administrative proceedings for environmental issues towards realization of sustainable development through better connection between the governed and those who govern.[94] The Aarhus Convention (of which the EU is a signatory)[95] stipulates parties must make provision for public participation in decisions on specific environmental activities; in plans, programmes and policies relating to the environment; as well as in preparation of executive regulation and

---

[91] European Commission, 'Life Sciences: A Strategy for Europe' COM (2002) 27 final, 23 January 2002, 28.

[92] Despite its strong endorsement of wide participation, the White Paper on European Governance precluded the employment of legal rule, due to 'excessive rigidity and risk slowing the adoption of particular policies' (n. 89) 17. Galligan found this reasoning unconvincing since the claim that such an approach would lead to rigidity is bluntly asserted without the slightest concern for any supporting evidence, while the wisdom of a legal approach that may complement other approaches to participation is not even considered. 'The White Paper plainly lost an opportunity to formulate a rigorous, properly researched approach to public participation in governance'; Denis Galligan, 'Citizen's rights and biotechnology regulation' in Francesco Francioni (ed.), *Biotechnologies and International Human Rights* (Hart Publishing 2007) 335, 350.

[93] Galligan (n. 92) 351.

[94] Principle 10 of the Rio Declaration on Environment and Development 1992, (1992) 31 *International Legal Materials* 874.

[95] Council Decision of 17 February 2005 on the conclusion, on behalf of the European Community, of the Convention on access to information, public participation in decision-making and access to justice in environmental matters (2005/370/EC) OJ L 124/1. Public participation in decision-making is identified as one of the three pillars of the Aarhus Convention, the others being public access to information and public access to justice. See further, Maria Lee, 'Public participation, procedure, and democratic deficit in EC environmental law' (2004) 3 *Yearbook of European Environmental Law* 193.

generally applicable legally binding normative instruments.[96] The Almaty amendment to the Aarhus Convention specifically requires Members to introduce an additional regime for public participation in decisions on the deliberate release into the environment and placing on the market of GMOs.[97] This amendment stipulates an early and effective public participatory measure that is prior to making specific decisions on the release of GMOs, and underlined that 'due account is taken of the outcome of the public participation procedure'.[98]

An argument for participation, that is, for mechanisms that allow the public to meaningfully influence regulatory decisions, at a broad policy level or in respect of individual releases of GMOs – could be criticized for the institutional costs incurred and the (im)practicalities involved. Prescriptions of how to create such organizational structures within public bodies are not envisaged as part of this book; such institutional and organizational questions are not germane to the central question of this book. Within its mandate of examining the extent of legal effectuation of public participation in GMO regulation, this book seeks to carefully analyse the relevant provisions to see whether they effectuate participation, and, further, whether the regulatory tools have conceptual possibilities for law to incorporate such participatory requirements.

How participation can be structured is intricately related to who is recognized as capable of participation, and how that public is conceptualized. Lee and Abbot focus on a serious limitation of the Aarhus

---

[96] See Articles 6, 7 and 8 of Aarhus Convention. Article 6 (8) requires Members to take due account of the outcome of the public participation in decisions on specific activities. Article 7 mandates parties to provide opportunities for public participation in the preparation of policies relating to the environment, to the extent appropriate. Article 8 requires parties to strive for promoting effective public participation at an appropriate stage, and while options are still open, during the preparation by public authorities of executive regulations and other generally applicable legally binding rules that may have a significant effect on the environment. It stipulates Members must take the result of the public participation into account, as far as possible.

[97] Decision II/1 on genetically modified organisms, Amendment to the UNECE Convention on access to information, public participation in decision-making and access to justice in environmental matters adopted at the second meeting of the Parties held in Almaty, Kazakhstan, on 25–27 May 2005, UN ECOSOC ECE/MP.PP/2005/2/Add.2 20 June 2005, <http://www.unece.org/index.php?eID=tx_nawsecuredl&u=0&filefileadmin/DAM/env/documents/2005/pp/ece/ece.mp.pp.2005.2.add.2.e.pdf&t=1322648754&hash=06f4d8cfd8f65640718ab1c2f47a43a204d92566> last accessed 20 November 2011.

[98] Annex I bis (7) ibid.

Convention regarding the real nature of participation.[99] They observe the relative privileging of NGOs in the Aarhus Convention:

> [E]ven if environmental interest groups represent one particular view of the public interest, arguably creating a public interest in their involvement, they cannot claim to represent the public ... What we can say is that since industry or developers are undoubtedly involved in any decision making process, environmental interest groups provide an invaluable alternative input, particularly since negotiation with the regulated industry is the starting point for much regulatory reform.[100]

The public sphere can be imagined as a public zone of mediation between the state and individuals or groups, where various publics can participate in state processes of decision-making through negotiations, contestations and deliberations. Habermas' work on the public sphere is widely recognized as the primary reference point in understanding conceptualizations of the public sphere today, even by its critics who see it as inadequate and erroneous.[101] Gardiner succinctly summarizes the central thesis of Habermas' work:

> [I]n [the] eighteenth and nineteenth centuries, a distinct forum for rational public debate emerged in most Western European countries. It constituted an area of social life, separate from the state apparatus, in which citizens

---

[99]   Lee and Abbot (n. 86) 107.

[100]   Ibid 87.

[101]   Jurgen Habermas, *The Structural Transformation of the Public Sphere* (Polity 1989). Significant work on the public sphere already existed when Habermas originally wrote his book in German in 1962, for instance of Bakhtin and Dewey (cited later in this footnote). The ideal of rational and informed discussion of public policy is said to be one which runs like a red thread through the whole of Habermas' work – William Outhwaite, *Habermas: A Critical Introduction* (Stanford University Press 1994) 8. The Russian philosopher Bakhtin historicized the public sphere of eighteenth century Europe, like Habermas. In contrast, however, Bakhtin suggested that there was never a superseding of social antagonisms through sober and virtuous debate as a vehicle of rational consensus: Mikhail Bakhtin, *The Dialogic Imagination: Four Essays by M. M. Bakhtin* (Michael Holquist ed., Michael Holquist and Caryl Emerson trs, Texas University Press 1981). Habermas later clarified that historically a generalized commitment to collective and rational self-determination was neither present nor realized, and conceded that the public sphere in his conceptualization, more or less, comprised well-educated bourgeois males: Habermas (n. 106) 48. See also John Dewey, *The Public and its Problems* (Holt 1927). See further, Andreas Koller, 'The public sphere and comparative historical research: An introduction' (2010) 34 *Social Science History* 261.

gathered to converse about the issues of the day in a free and unrestricted fashion, either literally, as in the town square, or in the pages of diverse journals and periodicals. Debate proceeded according to universal standards of critical reason and argumentative structure that all could recognize and assent to; appeals to traditional dogmas, or to arbitrary subjective prejudices, were ruled inadmissible. Thus it was in the public sphere that 'discursive will formation' was actualized in a manner that represented *general* social interest, as opposed to a class or sectional one.[102]

The Habermasian public sphere also sought to be a historical account of eighteenth–nineteenth century Germany, Britain and France, where 'the critical reasoning of the public constitutes an effective steering force in both society and polity'.[103] An approximation of the male domain as the bourgeoisie public sphere was central in such an account, 'founded upon free and equal access and upon willing consent between participants'.[104] The neglect of public spheres other than the male bourgeois public sphere in his approximation has been criticized for overlooking the 'coercive and power-driven attributes of sectionalism, exclusiveness and repression', 'having profound consequences, not only for historical and social investigation but also for theoretical speculation'.[105] Habermas subsequently conceded the existence of a plurality of publics:

> [T]he modern public sphere now comprises several arenas in which ... a conflict of opinions is fought out more or less discursively. This conflict does not merely involve a competition among various parties of loosely associated private people; from the beginning a dominant bourgeois public *collides* with a plebeian one.[106]

---

[102] Michael E. Gardiner, 'Wild publics and grotesque symposiums: Habermas and Bakhtin on dialogue, everyday life and public sphere' in Nick Crossley and John Michael Roberts (eds), *After Habermas: New Perspectives on the Public Sphere* (Blackwell 2004) 28.

[103] John Michael Roberts and Nick Crossley, 'Introduction' in Nick Crossley and John Michael Roberts (eds), *After Habermas: New Perspectives on the Public Sphere* (Blackwell 2004) 1.

[104] See Geoff Eley, 'Nations, publics, and political cultures: Placing Habermas in the nineteenth century' in Craig Calhoun (ed.), *Habermas and the Public Sphere* (MIT Press 1996) 321.

[105] Roberts and Crossley (n. 103) 11–12. See further, ibid; Janet *Siltanen* and Michelle *Stanworth,* 'The politics of private woman and public man' in Janet *Siltanen* and Michelle *Stanworth* (eds), *Women and the Public Sphere: A Critique of Sociology and Politics* (Hutchinson 1984).

[106] Jurgen Habermas, *Moral Consciousness and Communicative Action* (Polity 1992) 430. Emphasis added. Some theorists extend this to an argument that the bourgeois public sphere itself arose as a response to the ambivalent,

This concept of multiple publics is developed by feminists like Nancy Frazer, and others like Craig Calhoun, to argue for recognition of the legitimate discursive claims of those residing in 'alternative public spheres', or counterpublics that are 'parallel discursive arenas where members of subordinated social groups invent and circulate counter discourses'.[107] Frazer points to the fallacy in the idea that inequalities between participants can be bracketed during discursive deliberation, when in fact it only 'conceals real inequalities' including access to resources, which can have 'drastic consequences for the outcome of debate and discussion'.[108] She points out that fixed boundaries on topics of public discussion that are structured around a common interest do not exist a priori, but are products of discourse and dialogue. Formation of counterpublics permits 'subordinated groups to formulate oppositional interpretations of their identities, interests and needs'.[109] She therefore argues that a participatory parity in the public sphere requires 'the elimination of systematic social inequalities', and not merely its bracketing, and in situations where such inequality persists it is preferable to construe a 'multiplicity of mutually contestatory publics' as opposed to a 'single modern public sphere oriented solely to deliberation'.[110]

Following this, the book seeks to investigate if and how existing policy commitments to public participation have been transposed to legal justiciability. The choice of public participation as a referral point in this investigation is notwithstanding the acknowledgement of the problems posed by heuristics of public participation, viz., the real possibility that public participation may 'simply hold a mirror up to the pattern of power in the community; if the rich and well-organized are heard, while the

---

expressive and effectual practices of the Other – Michel de Certeau, *The Practice of Everyday Life* (University of California Press 1984); and further 'actively suppresses sociocultural diversity in constituting an arena inimical to difference' – Robert Asen, 'Seeking the "counter" in counterpublics' (2000) 10 *Communication Theory* 424, 425.

[107] Nancy Fraser, 'Politics, culture, and the public sphere: Toward a postmodern conception' in Linda J. Nicholson and Steven Seidman (eds), *Social Postmodernism: Beyond Identity Politics* (Cambridge University Press 1995) 291, 295. See further, Michael Warner, 'Publics and counterpublics' (2002) 88 *Quarterly Journal of Speech* 413; Roberts and Crossely (n. 103) 14; Craig Calhoun, 'The public sphere in the field of power' (2010) 34 *Social Science History* 30.

[108] Fraser (n. 107) 291.

[109] Ibid 116.

[110] Ibid 295.

poor and minorities are weakly represented'.[111] Nevertheless, it seeks to investigate the possible (in)consistencies between the claim to participation in EU policy documents and its transposition in EU GMO regulation through legal and justiciable principles. This is with an ambition that such a comparison can spark a genuine shift as a result of which patterns of power and productions of rationality will also be questioned. A legal engagement with public concerns about EU GMO regulation is required for factors identified earlier in this chapter: a) normative aspirations for further democratization of law to be proximate to public preferences; b) in policy areas involving perceptible, considerable and persistent public disquiet; c) especially in a polity where concerns about the democratic deficit are repeatedly raised; and d) consequentialist advantages of public participation for development of socially appropriate technology. Recognition of the legitimate issues of power imbalances and regulatory capture (that can beset both classical regulatory mechanisms and public participatory mechanisms) need not be normatively debilitating for the investigation in this book, but can be taken as an important factor that needs cognizance and additional attention.

The investigation in this book regarding existing and possible room in the EU regulation of GMOs for participation of publics outside of techno-scientific communities moves beyond a review of relevant legislation. As Glowka noted: 'General references to public participation may not translate into actual public participation if additional criteria are not provided on the form that public participation can take'.[112] After providing a broad description of legislation that frames the EU regulation of GMOs in Chapter 2, including labelling, coexistence and liability, the book identifies safety as the central element in the regulatory framework for release of GMOs. The concept of risk and the principle of precaution are seen as two important motifs in the regulatory pursuit of safety. Later chapters undertake a conceptual exploration of these two motifs to examine if and how notions of participation of publics, which are wider than expert scientific communities, are conceptually central in them.

The third chapter examines the concept of risk as a regulatory tool to pursue safety, and is followed by an exploration of the nature and

---

[111] Lee and Abbot (n. 86) 107.

[112] Lyle Glowka, *Law and Modern Biotechnology: Selected Issues of Relevance to Food and Agriculture* (FAO legislative study 78, FAO 2003) 51 <ftp://ftp.fao.org/docrep/fao/006/y4839E/y4839E00.pdf> last accessed 12 January 2011.

implementation of the precautionary principle in Chapter 4. The conceptual possibilities of bringing wider public concerns within risk and
precaution are examined. The fifth chapter describes the global rules
related to the regulation of GMOs in the GATT/WTO framework and the
Cartagena Protocol to the Convention on Biodiversity. Through this, the
chapter examines whether these global rules provide room for possible
improvements of EU safety regulation that include public participation.
This chapter also includes a short section that describes the manner in
which the WTO framework recognizes Members' regulatory action to
protect public values. Chapter 6 investigates different regulatory strategies employed for protecting public values related to GMOs in the EU.
Continuing from the description of labelling and coexistence from
Chapter 2, the sixth chapter elaborates on the purpose and limitations of
furthering the regulatory recognition of ethical plurality as a strategy to
pursue public values about GMOs. The chapter analyses the nature and
quality of participation that is possible through consumer decisions in the
regulation of research, development and use of GMOs. It further examines the use of public bioethics committee reports as a marker of public
values regarding the regulation of GMOs. This examination is towards
analysing the room made available for public participation in EU
regulation of GMOs through such reports. In conclusion, the ambition of
the book is restated for the sake of clarity. It describes relevant EU laws
to identify and analyse the room available or possible for public
participation in the EU regulation of GMOs. The first step for such a
project is a cursory description of the EU legal framework for GMO
regulation, attempted in the next chapter.

2. The EU legal framework for GMO regulation

> How we think about disagreements will determine how we think about politics. And since law is the offspring of politics, how we think about disagreement will determine in some measure how we think about positive law.[113]

EU law regulates the development and use of GMOs through allocating responsibilities to different authorities, public and private, accompanied by limited recognition of rights of public information, consultation and participation. As elaborated later in the chapter, a case-by-case examination of the risks of new GMOs, which are sought to be released for cultivation or other market purposes, is slotted between restrictions on contained use for research, and post-release specifications like labelling, traceability, coexistence and liability. This places GMO regulation within the interstices of EU environmental law, consumer law, agricultural law, food safety law and, of course, general internal market law.[114] This chapter attempts a cursory description of EU legislation that provides the

---

[113] Waldron (n. 57) 36.

[114] Article 4 (2) TFEU lists the competence regarding internal market, agriculture, environment, consumer protection and 'common safety concerns in public health matters' as shared between the Union and the Member States. Article 11 TFEU stipulates requirements of environmental protection to be integrated into the definition and implementation of the Union's policies and activities towards promotion of sustainable development. Requirements of consumer protection are also expected to be taken into account in defining and implementing other Union policies and requirements (Article 12 TFEU). Article 38 TFEU requires the application of internal market rules to agricultural products (with certain exceptions), and must be accompanied by a common agricultural policy that shall seek inter alia to increase agricultural productivity by promoting technological progress [Article 39 (1) (a)], and to ensure a fair standard of living for the agricultural community [Article 39 (1) (b)], taking into account the particular nature of agricultural activity, resulting from the social structure of agriculture and from structural and natural disparities between the various agricultural regions [Article 39 (2)].

framework for EU regulation of GMOs, and to identify the room provided for public participation, consultation and information in such regulation.

The EU accession to the European Convention on Human Rights (ECHR) makes EU regulation of science and technology, including GMOs, subject to the general rights and protection offered by the Convention.[115] The European Charter of Fundamental Rights, which emphasizes that the Union is founded on the indivisible and universal values of human dignity, freedom, equality and solidarity, and on the principles of democracy and the rule of law, is recognized as part of the broad ethical framework in the EU regulation of research, development and use of science and technology.[116] Specific insistence on public participation in EU decision-making is mostly absent in the Treaty on European Union (TEU), the TFEU and the ECHR.[117] Representative democracy is signified in the TEU as the foundation of the EU, where the European Parliament, the Member States through their governments, and political parties at the European level are expected to represent and express the will of the citizens.[118] However, the TFEU insists on citizens' right of access to EU documents, and prescribes that EU institutions conduct their work as openly as possible 'in order to promote good governance and ensure the participation of civil society'.[119] An ingenious platform is provided by the TEU where

---

[115] *Convention for the Protection of Human Rights and Fundamental Freedoms, 1950*, 213 U.N.T.S. 221, ETS No. 5, <http://conventions.coe.int/treaty/Commun/QueVoulezVous.asp?NT=005&CL=ENG> last accessed on 5 December 2011. The accession of the EU to the ECHR is stipulated by Article 6 (2) TEU and facilitated by Article 59 (2) ECHR.

[116] European Group on Ethics in Science and New Technologies (EGE), *Opinion No. 25 on Ethics of Synthetic Biology* (EGE 2009) 35–36. See, the *Charter of Fundamental Rights of the European Union, 2000*, 2000/C 364/01, <http://www.europarl.europa.eu/charter/pdf/text_en.pdf> last accessed 24 November 2011.

[117] In contrast, Flear and Vakulenko unconvincingly argue that participation can be read into the TEU. They argue that references to the EU being founded on respect for fundamental freedoms [Article 6 (1) TEU] and to 'constitutional traditions common to the Member States' [Article 6 (2) TEU] allude to public participation. Mark Flear and Anastasia Vakulenko, 'A human rights perspective on citizen participation in the EU's governance of new technologies' (2010) 10 *Human Rights Law Review* 661, 671.

[118] Article 10 (1) TEU.

[119] Article 15 (1) and (3) TFEU.

not less than one million citizens who are nationals of a significant number of Member States may take the initiative of inviting the European Commission, within the framework of its powers, to submit any appropriate proposal on matters where citizens consider that a legal act of the Union is required for the purpose of implementing the Treaties.[120]

Apart from these provisions, there are few legislative suggestions for a move towards public participation in EU regulation of science and technology.

By examining the three stages of EU regulation of GMOs,[121] the picture that emerges in this chapter is that this regulatory labyrinth is dominated by the motif of safety, articulated through the concept of risk and the principle of precaution. The chapter discusses the procedures related to contained use and field trials of GMOs, during their development, before a research entity can apply for authorization of release. Further, it discusses the steps involved in arriving at a regulatory decision to authorize the release of GMOs, either as an agricultural crop, human food or animal feed (feed). It includes an identification of provisions that facilitate public participation, consultation and information during such regulation. In the post-release procedures, it further discusses EU instruments of environmental guarantees, the relevant product safeguard measures, labelling and traceability requirements, coexistence measures and liability provisions. The chapter does not attempt to provide a detailed analysis of the relevant legal rules in an exhaustive fashion, but merely introduces the broad contours of the framework to an unfamiliar reader. After such an introduction, further discussions regarding EU procedures related to safety, labelling, segregation of different streams of agriculture (GM, conventional and organic), and consideration of public values apart from safety, are continued in later chapters.[122]

The EU Regulation which applied the provisions for public access to information and justice, and public participation in decision-making in environmental matters in the Aarhus Convention to EU law, stipulates that effective opportunities for public participation during the preparation,

---

[120] Article 11 (4) TEU.

[121] These are: procedures for authorization for release or use in the market, procedures prior to it and regulatory requirements after a decision to release.

[122] In particular in Chapters 4 and 6.

modification or review of plans or programmes relating to the environment are to be provided by EU institutions through appropriate provisions at an early stage, when all options are still open.[123] EU
institutions are required to identify the public that is likely to be affected
or have an interest in relevant programmes. Further, this Regulation also
stipulates that these groups are provided with public access to requisite
information, the regulators receive any comments from them, and the
regulators take due account of these comments.[124]

The complicated regulatory maze in the current attempt at harmonization in EU GMO regulation, which is discussed later in this section,
includes provisions for the general public to access relevant documents,
and to be consulted. The enactment of the Deliberate Release Directive,
2001[125] and the subsequent introduction of the two Regulations (Food
and Feed Regulation and Labelling Regulation)[126] in July 2003 completed
an important stage that ended the de-facto moratorium; where various states
had refused to allow the circulation of GMOs that were authorized by the
EU regime.[127] The legal compatibility of the moratorium with the WTO
framework was suspect and necessitated the formal end of the moratorium.
The attempt during the inception of the current regime was to tackle
the social and environmental issues emanating from the release of GMOs
in a manner acceptable to international and internal audiences.[128] Twin

---

[123]   Article 9 (1) Regulation 1367/2006/EC of the European Parliament and
of the Council on the application of the provisions of the Aarhus Convention on
Access to Information, Public Participation in Decision-making and Access to
Justice in Environmental Matters to Community Institutions and Bodies [2006]
OJ L264/13.

[124]   Article 9 (2) (3) and (4) ibid.

[125]   Directive 2001/18/EC of the European Parliament and of the Council on
the Deliberate Release of Genetically Modified Organisms into the Environment
and Repealing Directive 90/220/EEC [2001] OJ L 106/1 (Deliberate Release
Dir.).

[126]   Regulation 1829/2003/EC of the European Parliament and of the Council
on Genetically Modified Food and Feed [2003] OJ L 268/1 (Food and Feed
Reg.), and Regulation 1830/2003/EC of the European Parliament and of the
Council concerning the Traceability and Labelling of Genetically Modified
Organisms and the Traceability of Food and Feed products produced from
Genetically Modified Organisms and amending Directive 2001/18/EC [2003] OJ
L354 (Labelling Reg.). Both Regulations came into force on 18 April 2004.

[127]   See n. 36.

[128]   See the statements of the Environment Commissioner, Margot Wallström
and the Health and Consumer Protection Commissioner, David Byrne, on the
occasion of the adoption of the new Regulations: 'it will reinforce our international credibility and will certainly help in building public confidence in new

tropes of risk and consumer protection were employed in the regulatory regime to achieve a single market in authorized GMOs, seeking the maintenance of a high level of environmental protection and human health in a precautionary manner.[129] This is because, once a GMO is authorized by the Commission as safe for its intended use after the stipulated risk ascertainment process,[130] and appropriately labelled,[131] the same can be grown, sold or consumed in any individual Member State.[132] The precautionary principle is often asserted as the fundamental principle upon which this regulatory framework is built. The objectives of the Deliberate Release Directive affirm that 'in accordance with the precautionary principle, the objective of this Directive is to approximate laws, regulations and administrative provisions of the Member States and to protect human health and environment'.[133] Salient features of the regulatory framework include:

---

technologies ... we address the most critical concerns of the public regarding the environmental and health effects of GMOs and enable the consumers to choose ... European consumers can now have confidence that any GM food or feed marketed in Europe has been subject to the most rigorous pre-marketing assessment in the world'; Commission Press release, 'European Legislative Framework for GMOs in Place' (IP/03/1056, 22 July 2003).

[129]   Preamble 56 Deliberate Release Dir. See further, Preamble 5, 19 and 40 of the Directive.

[130]   See Articles 13, 14 (2), (3) and 15 (1) of the Deliberate Release Dir. and Articles 4–7, 16–19 and 28 of the Food and Feed Reg.

[131]   See Articles 2, 13, 24 and 25 of the Food and Feed Reg., and Articles 4–6 of the Labelling Reg.

[132]   Article 22 Deliberate Release Dir.; albeit in the case of seeds, the authorized seeds need to undergo the prescribed procedure to be inscribed into the EU Common Catalogue of varieties of Agricultural Plant Species, before they can be legally planted and harvested in the Community. See Directive 2002/53/EC of the Council on the Common Catalogue of varieties of Agricultural Plant Species [2002] OJ L153/2 (Common Catalogue Dir.) [2002] OJ L193/1.

[133]   Article 1 Deliberate Release Dir. The emphasis on the precaution is reiterated in Recital 8 of the Directive: 'the precautionary principle has been taken into account in the drafting of this Directive and must be taken into account when implementing it'. The principle is also mentioned in the Labelling Regulation (Recital 3). Further, the stipulation in the Food and Feed Regulation (Recital 6) for a case-by-case evaluation of the socio-economic, environmental and health risks for every application of release of GMOs, as opposed to the doctrine of substantial equivalence, demonstrates the precautionary inclinations in the regulatory system. See for more on the process versus product debate: Robert Howse and Donald Regan, 'The product/process distinction: An illusory basis for disciplining "unilateralism" in trade policy' (2000) 11 *European Journal of International Law* 249.

- a detailed set of principles, which is required to be considered in environmental and health risk assessments,[134]
- a requirement that applicants perform their own risk assessment prior to submitting an application for approval of the GMO,[135]
- broadening of the relevant matters to be considered in assessing applications to include ethical concerns,[136] and the cumulative long-term effects of GMOs on human health and the environment,[137]
- an emphasis on public consultation and access to information regarding applications for release,[138]
- provisions to ensure adequate labelling, traceability and post-market surveillance requirements,[139]
- segregation measures to ensure coexistence of GM, conventional and organic agriculture,[140]

---

[134] Article 4 (1) and (3), Deliberate Release Dir. is key in identifying this set of principles. The former requires the Member States to ensure that all appropriate measures, in accordance with the precautionary principle, are undertaken to avoid adverse effects that might arise from the deliberate release of GMOs on human health and the environment. The latter obligates Member States, and in appropriate cases the Commission, to accurately assess the potential adverse effects on human health and the environment, that may occur directly or indirectly through gene transfer from GMOs, on a case-by-case basis, taking into account the environmental impact according to the nature of the organism introduced and the receiving environment. Article 13 (2) (a) also requires the applicant to provide relevant information which shall inter alia take into account the diversity of sites of use of the GMO. Article 4 (2) Deliberate Release Dir. specifies that the environmental risk assessment envisaged by the regulatory framework includes the evaluation of risks to human health and the environment that may be posed by the deliberate release or market placement of GMOs – be it direct or indirect, immediate or delayed.

[135] Article 6 Deliberate Release Dir. and Article 5 (3) (e) Food and Feed Reg.

[136] Recital 9 Deliberate Release Dir., 'respect for ethical principles recognized in a Member State is particularly important. Member States may take into consideration ethical aspects when GMOs are deliberately released or placed on the market as or in products'. See also: Recitals 57 and 58 Deliberate Release Dir., which emphasize the need to engage with ethical concerns before release of GMOs. See further: Article 29 Deliberate Release Dir. and Article 33 Food and Feed Reg.

[137] Annex II, III A and III B, and Recital 19 Deliberate Release Dir.

[138] Recital 10 Deliberate Release Dir.

[139] See Article 21 Deliberate Release Dir., Articles 12, 13, 24 and 25 of the Food and Feed Reg., and Article 4 (6) of the Labelling Reg.

[140] Article 26a Deliberate Release Dir.

- a recent proposal to allow Member States measures that restrict or prohibit cultivation of authorized GM seeds, on specified grounds that are unrelated to the assessment of the adverse effect on health and environment,[141] and
- continuing attempts to create EU environmental liability regimes that are specifically applicable to GMOs.

2.1 AUTHORIZATION PROCEDURES UNDER THE DIRECTIVE AND REGULATION

The decision to release a GMO for agriculture or other purposes like animal feed, food, floriculture and food processing (the most visible stage of the regulation) is envisaged through an intricate mixture of the Deliberate Release Directive and the Food and Feed Regulation. There are partial overlaps and variations in the respective scopes of these two legal instruments. The methods underlying both the procedures are similar, viz., a characterization of the risks involved to the environment, health or both, accompanied by an attempt to gain political acceptability for this characterization.[142] However, there are significant differences between the formal procedures under the Directive and the Regulation, including the role of the Member State Competent Authorities.[143] This section describes the respective procedures under the Directive and the Regulation.

The Food and Feed Regulation is applicable in authorization of GMOs intended for a) use as food or feed, b) food or feed containing or consisting of GMOs, and c) food or feed produced from or containing

---

[141] Communication from the Commission to the European Parliament, the Council, the Economic and Social Committee and the Committee of the Regions on the Freedom for Member States to Decide on the Cultivation of Genetically Modified Crops, COM (2010) 380 final. The proposal pertains to reaching a decision to ban cultivation of GMOs on social, ethical and economic grounds, through an introduction of Article 26b in the Directive, 'as long as they are in conformity with the Treaties', and are not violating WTO obligations.

[142] Preamble 9 of the Food and Feed Reg. specifies that the authorization procedures for food safety should include the principles introduced in the Deliberate Release Dir.

[143] See Maria Lee, *EU Environmental Law: Challenges, Change and Decision-making* (Hart Publishing 2005) ch. 9, for a detailed discussion on the difference between the requirements prior to a decision to release under the two instruments. Only a cursory description of the authorization is attempted here, partly since there is so much literature already available on this.

ingredients produced from GMOs.[144] The Court has held that products such as honey and food supplements containing genetically modified DNA ought to fall within the latter category of 'food produced from or containing ingredients produced from GMOs', even though such genetically modified DNA is incapable of reproduction or transfer of genetic material.[145] The Deliberate Release Directive is applicable to releases of GMOs into the environment for (a) any other purpose than for placing in the market, and (b) placing in the common market GMOs 'as or in products'.[146] There are significant overlaps between this latter category in the Directive and the scope of the Regulation, viz., GMOs for food or feed, and food or feed containing GMOs. In such cases of overlap, the application is processed under the Regulation for food safety, coupled with an environmental risk analysis under the Directive.[147] In situations like transgenic floriculture (having no applications of food), only the procedure under the Directive is applicable. In other cases of release that do not include living material, say of processed transgenic food, the application comes under the exclusive ambit of the Regulation since it involves no apparent environmental hazards.

Under both instruments, the applicant who is developing the GMO is expected to initiate the process of risk regulation for an authorization of release by forwarding a food safety assessment, an environmental risk assessment, or both (in applicable cases). Any application to authorize release of a GMO is to include this assessment in the dossier, and is to be submitted to the relevant Member State Competent Authority.[148] At this initial stage, apart from undertaking a preliminary evaluation of the hazards, the applicant has responsibility to share all available peer reviewed and independent studies regarding the safety of the GMO. If the application for release is made exclusively within the context of the Directive, the national authority to which the application was originally addressed examines the relevant environmental risks, including through

---

[144]   Articles 3 and 15 Food and Feed Reg. In the interest of readability, 'food' is used in the following text to include animal feed, unless specified otherwise.

[145]   *Karl Heinz Bablok* v *Freistaat Bayern* Case C-442/09, Judgment of the Court (Grand Chamber) 6 September 2011, not yet reported.

[146]   Article 1 Deliberate Release Dir.

[147]   Articles 6 (4) and 18 (4) Food and Feed Reg. requires that environmental safety procedures for food containing/consisting of GMOs applies as in the Deliberate Release Dir.

[148]   Article 4 (2) Deliberate Release Dir., Article 5 (3) (e) and Article 17 (3) (e) Food and Feed Reg.

an appraisal of the dossier submitted by the applicant. The environmental risk assessment by the specific Authority is circulated to all other Member States' Competent Authorities and the Commission within a stipulated time. In case a recommendation to authorize release is made in this report, other Member States and the Commission are granted an opportunity to provide comments or reasoned objections to the recommended release.[149] The Commission is expected to make the summary of the dossier of the original application and the national Competent Authority's assessment report available to the public. It must also forward any consequent comments by the public to the relevant Competent Authorities within the stipulated time.[150]

If there are no objections from scientific authorities of other Member States, or if the objections raised are resolved through exchange of positions within the stipulated time, the relevant GMO is authorized to be released under the Directive.[151] Although the regulatory structure provides this early opportunity to arrive at an agreement on outstanding issues between the national authorities, such a resolution is yet to occur under this regime. In these instances of disagreement between national scientific authorities, the Commission is required to forward the dossier to the Scientific Committee for consultations (the GMO panel of the European Food Safety Authority (EFSA) is the designated committee for this purpose).[152] Based on the report of EFSA as the Scientific Committee, the Commission is expected to forward a draft decision on the possible release for a general EU procedure called Comitology.[153] In this committee procedure, expert representatives of various Member States deliberate and vote on the proposal.[154]

The initial stages in the regulatory procedure for authorization are substantially different under the Regulation. Here Member State Competent Authorities have minimal roles, though individual applications are

---

[149]   Article 15 (1) Deliberate Release Dir.

[150]   Article 24 (1) ibid. Further, the Member States have the responsibility to apply principles of Regulation 1049/2001 of the European Parliament and of the Council regarding public access to European Parliament, Council and Commission documents [2001] *L 145/61*, for applications regarding public access in this regard.

[151]   Article 11 (2) Deliberate Release Dir.

[152]   Article 28 ibid.

[153]   Ellen Vos, 'The role of Comitology in European governance' in Deirdre Curtin and Ramses Wessel (eds), *Good Governance and the European Union: Concept, Implications and Applications* (Intersentia 2005) 107.

[154]   Qua a joint reading of Articles 18 (1) and 30 (2) of the Deliberate Release Dir.

formally submitted to them. Applications under the Regulation are forwarded from these authorities within a stipulated time to EFSA, which circulates the same to other national authorities and the Commission. There is also an important provision to make the summary of the dossier available to the public.[155] Subsequent to this, EFSA carries out a safety assessment, which may include a number of opportunities for national authorities to be consulted.[156] This risk assessment of EFSA is submitted to the Commission, and is required to be made public so that citizens can make comments to the Commission regarding the contents of the report.[157]

The Commission is supposed to prepare a draft proposal based on the opinion of EFSA, any relevant provisions of EU law and 'other legitimate factors relevant to the matter'.[158] Though there is no explicit reference to the consideration by the Commission of public comments about EFSA opinion during the preparation of its proposal, this can be implied as part of 'other legitimate factors'. Here, the Commission is required to provide an explanation for any difference between the opinion of EFSA and its draft proposal regarding the application.[159] The Commission's draft proposal regarding the release of the specific GMO is required to be forwarded to the Standing Committee on the Food Chain and Animal Health for Committee procedure.[160]

---

[155]   Article 29 (1) of the Food and Feed Reg. Also, Article 28 of the Food and Feed Reg. requires the Commission to establish a Community register of GM food and feed that shall be made available to the public.

[156]   Articles 6 and 18 Food and Feed Reg. For instance, in case of an overlap with the Deliberate Release Dir., EFSA is required to consult all the Member State Competent Authorities, and in specific cases it shall ask a national authority of its choice to carry out an environmental risk assessment. EPEC Report to DG Sanco, *Evaluation of the EU legislative Framework in the Field of Cultivation of GMOs under Directive 2001/18/EC and Regulation (EC) No 1829/2003 and the placing on the market of GMOs on or in products Under Directive 2001/18; Final Report* (2010) found 'broad acceptance that it would be helpful to widen [Member State] participation in the risk assessment process and to bring greater expertise to bear on the appraisal of environmental risk assessments conducted by Member State authorities and EFSA', p. 20.

[157]   Articles 6 (7) and 18 (7) ibid.

[158]   Articles 7 (1) and 19 (1) ibid.

[159]   Ibid. 'The legal and political incentives to justify all decisions on a scientific basis are not likely to be overcome by simply adding on "other legitimate factors"': Maria Lee, 'Multi-level governance of GMOs in the European Union: Ambiguity and hierarchy' in Bodiguel and Cardwell (n. 28) 101, 105.

[160]   Article 35 (1) Food and Feed Reg.

The Committee procedures provided under the Directive and the Regulation are similar, and have invariably resulted in deadlocks, where neither the opposition nor the supporters of the proposal to release could garner the requisite qualified majority of two-thirds of the members of the Committee.[161] In these situations of deadlock in the Committee, the dossier is forwarded to the Council, with the assumption that in view of the contentious nature of the dossier, political responsibility needs to be explicitly exercised by the members of the Council. This is an expression of the bifurcation of EU risk regulation into technical risk assessment and political management, examined in detail in the next chapter. The voting pattern and the deadlock during voting in Regulatory Committee is mirrored in the Council votes, demonstrating the deep divisions and concerns that have continued to underpin decisions on most releases of GMOs in the EU.[162] Every application for release of GM crops has undergone the regulatory rigmarole of first, an assessment either by EFSA, a Member State Competent Authority, or both, and secondly, risk management considerations including the Committee procedure. Such a long chain of decision-making has not resulted in a qualified majority either to accept or reject the proposal. All these applications have been authorized by the Commission on the legimating factor of EFSA opinion, either as a risk assessor under the Regulation or as Scientific Committee that is consulted during committee procedure under the Directive.[163] The implications of this scenario for governance of controversial technologies

---

[161] Qua Article 5 and 7 of Council Decision 1999/468/EC of 28 June 1999 laying down the procedures for the exercise of implementing powers conferred on the Commission, and Article 205 (2) of the EC Treaty 29.12.2006 OJ C 321 E/1. Article 35 (2) of the Food and Feed Reg. and Article 30 (2) of the Deliberate Release Dir. require decisions to be made based on Articles 5 and 7 of Council Decision 1999/468/EC. See n. 167 for more on deadlocks.

[162] Lee, 'Ambiguity and Hierarchy' (n. 159) 110. See: 'EU meeting on GM maize ends in deadlock', 26 February 2009 <http://www.euractiv.com/sustainability/eu-meeting-gm-maize-ends-deadlock/article-179795> last accessed 1 December 2011; Council of the European Union, 2849th council meeting Agriculture and Fisheries, 18 February 2008, Press Release 6199/08, <http://www.consilium.europa.eu/ueDocs/cms_Data/docs/pressData/en/agricult/98819.pdf> last accessed 1 December 2011; 'GMO debate continues to divide EU', <http://www.euractiv.com/climate-environment/gmo-debate-continues-to-divide-eu/article-170387> last accessed 1 December 2011. See for more recent details <http://www.gmo-compass.org/eng/gmo/db/> last accessed 1 December 2011, and <http://ec.europa.eu/food/dyna/gm_register/gm_register_auth.cfm?pr_id=5> last accessed 1 December 2011.

[163] Ibid. All decisions can be found in the Eurlex database.

in general, and for risk regulation in particular, are discussed in the next chapter. However, the legitimacy of the Commission's decisions to authorize specific GMOs when misgivings on risk are unresolved, and where a simple majority in the Scientific Committee opposes the decision during bifurcated risk regulation, needs further reflection.

The recent change in rules regarding the Commission's exercise of implementing powers is relevant for both the Committee procedure and voting in the Council. The new requirement to arrive at a decision on the Commission's proposal of implementation is a double majority, that is, 'a qualified majority of 55% of the members, comprising at least fifteen of them and representing Member States comprising at least 65% of the population of the Union'.[164] Further, in situations where no opinion on the draft implementing acts related to protection of health or safety of humans, animals or plants is delivered, the Commission is expected to either resubmit an amended version of the proposal to the same Committee, or forward the proposal to an appeal committee, within the stipulated time.[165] In case a deadlock (no opinion) continues at this stage of appeal, the Commission is again given the discretion to adopt its own proposal to release.[166] However, these changes to the Committee procedure, post-Lisbon, have not altered the record of deadlock, where neither the Committee nor the Appeal Committee have been able to deliver an opinion till date.[167]

2.2 REGULATION PRIOR TO AUTHORIZATIONS FOR PLACING ON THE MARKET

The consideration of risks of GMOs does not, of course, begin with the aforementioned stage of authorization, but is preceded by a number of

---

[164] Qua a joint reading of Article 16 (4) TEU, and Article 5 (1) of Regulation (EU) No. 182/2011 of the European Parliament and of the Council of 16 February 2011 laying down the rules and general principles concerning mechanisms for control by Member States of the Commission's exercise of implementing powers.

[165] Article 5 (4) ibid (Regulation).

[166] Article 6 (3) ibid.

[167] Decisions on authorizations post the application of Lisbon Treaty in 2012 can be found on the GMO register, http://ec.europa.eu/food/dyna/gm_register/index_en.cfm.

regulatory measures that are mentioned in this section.[168] The Contained Use Directive provides guidelines for safe use during relevant research and development within the confines of laboratories and greenhouses.[169] Though the scope of the Directive is limited to microorganisms, some Member States have extended national implementation to contained use of all GMOs.[170] A prior assessment of the risk to health and environment from the research concerned must be carried out by the user of a modified organism; this is required to be made available to the national Competent Authority in appropriate forms.[171] Different categories provided under the Contained Use Directive give rise to different levels of containment of the specific activity, as well as differential obligations of notifications to the concerned public authority. The user is responsible for classifying the GMO sought to be used within containment into one of the four statutorily enumerated categories, viz., activities of no or negligible risk, activities of low risk, activities of moderate risk and activities of high risk, based on their own risk assessment.[172] The higher levels have more stringent containment and isolation conditions, accompanied by requirements of further periodic notifications to the Competent Authorities.[173] The Competent Authorities shall examine the conformity of the notifications with the requirements of this Directive, including the accuracy and completeness of the information given, the correctness of the risk assessment, the selection of the class of contained use and, where appropriate, the suitability of the containment and other protective measures, the waste management and emergency response measures.[174] In appropriate situations the Member States are allowed to provide for public consultations on aspects of the proposed contained use that they consider appropriate.[175]

---

[168] Though it might have been logical in this chapter to begin with a description of these measures, a prior examination of the later stage of authorization helps articulate the persistence of serious disagreements sharply.

[169] Directive 2009/41/EC of the European Parliament and of the Council on the contained use of Genetically Modified micro-organisms [2009] L 125/75 (Contained Use Dir.).

[170] See for instance the case of Wales. Cited in Marine Friant-Perrot, 'The EU regulatory regime for GMOs and its integration into Community food law and policy', in Bodiguel and Cardwell (n. 28) 79, 82.

[171] Article 4 of the Contained Use Dir.

[172] Article 4 and Annex III, sections A and B ibid.

[173] See Annex IV ibid.

[174] Articles 6–9 ibid.

[175] Article 12 ibid.

The Deliberate Release Directive provides guidelines for field trials of plants in ecosystems that could be affected by their use; such field trials are mandatory before placing such GMOs on the market.[176] The key document in a decision to authorize field trials is the technical dossier prepared by the applicant, and is similar in nature to the dossier used during the authorization procedures for both market release and contained use (mentioned earlier). The dossier is expected to supply information relating to GMOs, conditions of release and the 'potential receiving environment', interactions between the GMOs and the environment, control, remediation methods, waste treatment and emergency response plans as well as monitoring plans for identifying effects on human health or environment.[177] Based on the information provided in the dossier, the national authority is required to take a decision on the authorization of field trials. The Directive facilitates circulation of this information between various national authorities through the Commission,[178] and information on all field trial releases is made accessible to the public within a reasonable period, including exchanges between the national authorities.[179] Further, the public must be consulted with regard to these experimental releases.[180]

It was noted in the previous section that concerns about safety dominate the process to authorize deliberate release of GMOs, where political, ethical and other social concerns articulated outside scientific rationality appear to be underrepresented in law's engagement with contestations about GMOs as an agri-technological trajectory (argued in Chapter 4). There are, however, statutory provisions to institutionally acknowledge general concerns about ethics of the development and use of GMOs, through consultation with expert committees (examined in Chapter 6). On the other hand, concerns about safety continue beyond the stage of authorization, despite being raised before the inception of this stage, and are described in the next section.

---

[176] Preamble 25 Deliberate Release Dir.
[177] Article 6 (2) ibid.
[178] Article 11 ibid.
[179] Article 9 (2) ibid.
[180] Article 9 (1) ibid.

2.3 PERSISTENCE OF SAFETY CONCERNS BEYOND THE DECISION TO RELEASE

Concerns about safety that persist after authorization for commercial release of specific GMOs are addressed in two subsequent legal spaces. At a stage of regulation that is subsequent to the authorization for release, EU law provides certain powers to Member States to control movement and use of GMOs within their territories for responding to safety concerns about health and the environment. This section finds that although these two legal spaces have the potential to give expression to social disagreements about the risk assessed during authorization, scientific rationality strictly controls the application of these provisions.

2.3.1 Product Safeguard Measures

First among these areas is the combination of internal market law and the safeguard clause, seen in all EU product-based harmonizing measures.[181] A number of Member States have invoked these clauses to ban specific GMOs in their territories, including Austria, Hungary, France, Germany and Poland.[182] The Deliberate Release Directive empowers Member States to take derogatory action, subsequent to harmonization, with detailed 'scientific' grounds to demonstrate risk to human health or the environment, 'in the event of new or additional information' or 're-assessment of existing information on the basis of new or additional scientific knowledge'.[183] The absolute insistence on scientific evidence is different from safeguard provisions in the previous Deliberate Release Directive of 1990,[184] which had an additional ground. This ground allowed measures by Member States to provisionally derogate from the free movement of authorized GMOs, if there were 'justifiable reasons' to consider the constitution of risk to human health or the environment.[185] However, in the *Greenpeace* decision, where signatures of one-fifth of the population of Austria were submitted (along with certain scientific

---

[181] Article 114 TFEU explicitly authorizes these safeguard measures.
[182] See Margaret Rosso Grossman, 'Coexistence of genetically modified, conventional, and organic crops in the EU: The Community framework' in Bodiguel and Cardwell (n. 28) 123, 149.
[183] Article 23 of the Deliberate Release Dir.
[184] Council Directive 90/220/EEC of 23 April 1990 on the Deliberate Release into the environment of Genetically Modified Organisms [1990] OJ L17/15.
[185] Article 16 ibid.

documents) in support of the safeguard measure as constitutive of possible risk to their health, these were not regarded as 'justifiable reason'.[186]

The Common Seed Catalogue Directive,[187] an instrument that provides for a register of all seed varieties that are intended for cultivation in the EU (including GM seeds that are already approved by the authorization process under the Deliberate Release Directive) also provides for particular safeguard measures.[188] This Directive has safeguard provisions similar to the current Deliberate Release Directive and seeks to respond to urgent threats to the environment and health.[189] In addition, the Catalogue also permits safeguard measures when 'it is well-known that the variety is not suitable for cultivation in any part of its territory because of its type of maturity class'.[190]

The safeguard clause under the Food and Feed Regulation provides for urgent measures of suspension or termination of commercial releases in the event of a severe risk;[191] for example, in light of an EFSA opinion underlining an urgent need to suspend or modify an existing authorization.[192] The Court has recently emphasized that in light of the overall

---

[186]   Case C-6/99 *Association Greenpeace France and Others* v *Ministere de l'Agriculture et de la Peche and Others* [2000] ECR-I-1651.

[187]   Common Catalogue Dir. (n. 132). The Directive mandates registration in the catalogue as a seed which is 'distinct, stable and sufficiently uniform and … of satisfactory value for cultivation and use', Article 3.

[188]   Article 4 ibid.

[189]   Given the experience of interpretation of the Deliberate Release Dir. in *Austria* v *Commission* (n. 213) by the European Court of Justice (ECJ) (as it was then called), there is a high likelihood that a de-facto requirement of rigorous *scientific* proof could be read into these provisions under the Common Catalogue as well.

[190]   Article 16 (2) of the Common Seed Catalogue Dir. See further: Case C-165/08 *Commission* v *Poland* [2009] ECR I-6843.

[191]   Article 34 Food and Feed Regulation. The ECJ in *Monsanto Agricultura Italio SpA and Other* v*Presidenza del Consiglio del Ministri and Others*, Case C-238/01, [2003] ECR I-8105, held that the precautionary principle finds specific expression in the safeguard clause (para. 110). See for a detailed commentary of the case: Han Somsen and Thijs Etty, 'ECJ: Case report on Case C-236/01 *Monsanto Agricultura Italia Sp. A and Others*v *Presidenza Del Consiglio dei Ministeri and Others*' (2004) 13 *European Environmental Law Review* 3.

[192]   Articles 53 and 54 of Regulation 178/2002/ EC of the European Parliament and of the Council Laying down the General Principles and Requirements of Food Law, Establishing the European Food Safety Authority and laying down Procedures in the matters of Food Safety [2002] OJ L 31/1 (Food Safety Reg.). See the Judgment of the ECJ in the *Greenpeace* decision (n. 186) paras

scheme of the Regulation, the assessment and management of a serious and evident risk ultimately is the sole responsibility of the Commission and the Council, subject to review by the EU Courts.[193] Notwithstanding the temporary and preventive nature of these measures, the judgment reiterated that these measures may be adopted only based on a risk assessment that is as complete as possible in the particular circumstances of the case.[194] The Court of Justice of the European Union (CJEU) in *Monsanto (France)* held that the safeguard measures allowed through the Regulation had to refer to a '*significant* risk which clearly jeopardises human health, animal health or the environment', and 'must be established on the basis of new evidence based on reliable scientific data'. The Court emphasized that this insistence is notwithstanding their temporary character and preventive nature, and underlined that such measures may be adopted 'only if they are based on a risk assessment which is as complete as possible'.[195]

It is important that safeguard measures are generally expected to be provisional in nature, and require the concerned Member State to inform the Commission and other Member States about the action taken, including its reasons and a review of risk assessment.[196] Further, a subsequent centralized review process through the committee procedure of these measures (in most cases, tightly controlled by claims of scientific basis) ensues within the stipulated time.[197] Thus safeguard clauses under both the Deliberate Release Directive and Food and Feed Regulation not only harmonize and centralize short-term responses to even exceptional situations, but also tie their application to an unrealistically high bar of narrow scientific proof. This is well illustrated by the attempts of the Commission to repeal the Austrian safeguard measures that were in the form of a temporary ban on the use and sale of two strains of GM maize in its territory. The first two proposals by the Commission to withdraw the Austrian safeguard measures were rejected by the Council by a qualified majority. In the vote on the second proposal, the qualified majority pointed to the statutory requirement

---

71–74, regarding the relationship between these provisions and the safeguard provision in the Food and Feed regulation.

[193]   *Greenpeace* (n. 186) para. 78.

[194]   Ibid para. 77.

[195]   *Monsanto SAS and Others* v *Ministre de l'Agriculture et de la Peche* (joined C-58/10 to C-68/10P), not yet reported, [70] [76] [77].

[196]   Articles 23 and 30 of the Deliberate Release Dir., Article 34 of the Food and Feed Reg. and Article 54 of the Food Safety Reg.

[197]   Ibid.

under the Deliberate Release Directive to reassess the concerned GMOs, and argued that 'the different agricultural structures and regional ecological characteristics in the EU need to be taken into account in a more systematic manner in the environmental risk assessment of GMOs'.[198] The Commission's third proposal on the specific derogation, which incidentally targeted only the food and feed aspects of the safeguard action, ended in a deadlock. Here there was an inability to secure a qualified majority either in favour of or against the proposal, leaving the Commission with the obligation to adopt its own proposal to repeal the Austrian measures.[199] This situation elicited a response from the EU Presidency that most Member States had a second reasoning against the Commission's proposal to withdraw the safeguards, viz., 'the feeling that a member state's will should be respected'.[200] These events reveal a problematic situation, where the Commission in the name of implementing EU legislation has much leeway to act on its proposal (opposing specific Member State safeguard measures), notwithstanding significant majority support for these safeguards that falls short of a qualified majority in the Committee procedure. Such leeway for the Commission continues under the Committee procedure under post-Lisbon amendments.[201]

Significantly, the Commission has proposed an opt-out clause from allowing cultivation of GM seeds that were already authorized by EU regulation, in the form of an additional safeguard clause to the Deliberate Release Directive.[202] The Commission points out the lack of freedom for Member States to decide whether they wish to cultivate GMOs in their territory on the 'basis of specific conditions', due to the absence of a provision to take into account legitimate factors other than science during authorization of deliberate release under the Directive.[203] This proposal is subsequent to declarations from the Kingdom of Netherlands to the

---

[198]   See the minutes of the 2773rd Environmental Council Meeting, 18 December 2006, rejecting European Commission, 'Proposal for a Council Decision concerning the provisional prohibition of the use and sale in Austria of genetically modified maize' COM (2006) 510 final.

[199]   2826 Environmental Council meeting, 14th November 2007.

[200]   <http://www.euractiv.com/en/biotech/eu-considers-pause-thought-gmos/article-168053> last accessed 11 March 2011.

[201]   Text to n. 167.

[202]   See Communication from the Commission to the European Parliament, the Council, the economic and social committee and the Committee of the regions on the freedom for Member States to decide on the cultivation of genetically modified crops, COM (2010) 375 final.

[203]   Ibid 5–6.

Agriculture and Environmental Councils, asking the Commission to suggest a solution on cultivation while taking into account the socio-economic dimensions of GMO cultivation.[204] Further, Austria, supported by twelve Member States, presented a paper in the Environment Council suggesting an introduction of legislation that provides an opt-out clause for cultivation of specific GMOs.[205] Since cultivation has a strong local and regional dimension, the draft opt-out clause has sought to provide for a safeguard clause for Member States to restrict or prohibit cultivation on the basis of grounds relating to public interest other than protection of health and the environment.[206]

2.3.2 Environmental Guarantee Measures

A second space in the regulatory framework, where concerns of safety continue to be raised, is through environmental 'guarantee' measures. These guarantees empower a Member State to introduce national pro-visions after the adoption of a harmonization measure either by the European Parliament and Council, by the Council or by the Commis-sion.[207] New measures can be adopted by Member States to protect the 'environment or the working environment' on grounds of a problem specific to that Member State,[208] if it can be demonstrated to be based on 'new scientific evidence'.[209] Importantly, the environmental guarantee measures have a smaller range of potential triggers than the safeguard measures discussed in the previous section, since the safeguards can be

---

[204] Note with ref. 7581/09 of the EU Council, cited in ibid fn 10.

[205] Note with ref. 11226/2/09 REV 2 of the EU Council, cited in, ibid fn 12.

[206] European Commission, 'Proposal for a Regulation of the European Parliament and of the Council amending Directive 2001/18/EC as regards the possibility for the Member States to restrict or prohibit the cultivation of GMOs in their territory' COM (2010) 375 final, Article 1 and Recital 8.

[207] See Article 114 TFEU, which is the equivalent of Article 95 of the EC Treaty. It provides for qualified majority decision-making for internal market affairs, and was originally introduced to plug justifications of Member States vetoing Community proposals for the reason that they are outvoted and forced to reduce a national level of environmental protection to comply with an EC measure. See for more: Michael Doherty, 'The application of Article 95 (4)(6) of the EC Treaty: Is the Emperor still unclothed?' (2008) 8 *Yearbook of European Environmental Law* 48, 49.

[208] Article 114 (6) TFEU, upon which the procedure established under Article 114 (7), (8) and (9) TFEU shall be applicable.

[209] See for more on the difference between *maintenance* and *introduction* of new measures in this regard: Joanne Scott, 'European regulation of GMOs and the WTO', (2003) 9 *Columbia Journal of European Law* 213, 225–226.

invoked on the grounds of a serious risk either to human/animal health or the environment.[210] In contrast, environmental guarantees can be used in a more general manner; that is, states could notify measures that restrict the use of GMOs generally, as opposed to safeguard measures that can pertain only to specific GMOs, on a case-by-case basis. Unlike in safeguard measures, the Council has no role in deciding the legality of guarantee measures; this authority solely resides with the Commission and is subject only to review by EU courts.[211]

There are considerable limitations to the operation of these guarantee measures,[212] where the Commission's interpretation is that these national measures are only allowed if they are justified without endangering the unity of the internal market. The Commission is expected to approve these measures so as to avoid prejudice to the binding nature and uniform application of EU law, lest the establishment and functioning of the internal market be rendered ineffective by Member States retaining the unilateral right to apply derogatory national rules. These limitations are clearly emphasized in the judgment of the ECJ (as it was called then) in *Austria* v *Commission*.[213] The milieu of this case duly testifies to the continuing requirement of the centrality of science-based assessment for the applicability of Article 95 (5) of the EC Treaty (the predecessor and equivalent of Article 114 (5) of the TFEU) in a manner that is very similar to the restriction of safeguard provisions under the Deliberative

---

[210]   See Nicolas de Sadeleer, 'Procedures for derogations from the principle of approximation of laws under Article 95 EC' (2003) 40 *Common Market Law Review* 889, 890, where he emphatically states that the guarantee mechanisms are especially 'biased towards the protection of environment'.

[211]   Article 114 TFEU.

[212]   The Commission's interpretation is that these national measures are only allowed if they are justified without endangering the unity of the market: Communication of the European Commission concerning Article 95 (paragraphs 4, 5 and 6) of the Treaty of European Community COM (2002) 760, consideration 10.

[213]   Cases C-439/05 P and C-454/05 P *Land Oberösterreich and Republic of Austria* v *Commission*, [2007] *ECR* I-7141. This was an appeal from Judgment of the Court of First Instance (Fourth Chamber), 5 October 2005 in joined cases T-366/03 and T-235/04. The case concerns a ban of GMOs in Upper Austria for the protection of organic agriculture, nature and natural biodiversity, through the notification of the Upper Austrian Act on the Prohibition of Genetic Engineering, 2002, pursuant to Article 95 (5). These provisions are analogous with the procedure under Article 100a (4), introduced by the Single European Act of 1987. See: Case C-41/93 *France* v *Commission* [1994] ECR I-1829, paras 29 and 30, and Case C-319/97 *Kortas* [1999] ECR I-3143, para. 28, for cases under the old Article.

Release Directive. This is particularly so, because of the two requirements that (1) the Member State is expected to demonstrate that the notified measure was justified in the light of 'new scientific evidence' concerning protection of the environment; and (2) was also a problem 'specific' to the Member State.[214] Cogent critique exists that identifies the limitations and narrowness of the Court's approach in restrictively defining new scientific evidence to include only those data which were generated after the specific harmonization measure. This, it has been argued, has rendered the provision practically useless, and a more appropriate construction would have been 'to accept all relevant knowledge, including information that was already available at the time of adoption of the community measure, but which for some reason has been overlooked or deemed irrelevant at the time'.[215]

This experience in applying environmental guarantees in GMO regulation is similar to the general account of invocations of EU environmental guarantee provisions. Originally designed as a guarantee so that environmentally progressive Member States could retain or introduce high levels of protection after harmonization, these provisions have been more or less rendered redundant by restrictive interpretations of the de-facto requirement of new scientific evidence and specificity to privilege uniformity and the internal market.[216] No provisions exist for direct consultations or participation for eliciting public opinion regarding the invocation or continuation of such measures.

---

[214] Ibid (*Land Oberösterreich*) para. 44.

[215] See Floor Fleurke, 'What use for Article 95 (5) EC? An analysis of *Land Oberösterreich and Republic of Austria* v. *Commission*' (2008) 20 *Journal of Environmental Law* 267. See also the opinion of Advocate General Sharpston in *Austria* v *Commission*: 'I do not think it necessary to look further in order to reach the view that new conclusions drawn from existing data may constitute new scientific evidence within the meaning of Article 95 (5) EC'; ibid para. 124.

[216] A long list of critiques includes, Hanna Sevenster, 'The Environmental Guarantee after Amsterdam: Does the Emperor have new clothes?' (2000) 1 *Yearbook of European Environmental Law* 291; Roda Verheyen, 'The Environmental Guarantee in practice: A critique' (2000) 9 *Review of European Community and International Environmental Law* 179; Joakim Zander, 'The "Green Guarantee" in the EC Treaty: Two recent cases' (2004) 16 *Journal of Environmental Law* 65; Doherty (n. 207).

2.4   OTHER IMPORTANT ASPECTS OF EU REGULATION

An important aspect of post-release anxieties is captured in the legal complex of labelling, traceability and coexistence, which is placed within the larger context of food safety law and consumer law. Ensuring traceability and identification of the presence of GM in products has a number of objectives. These include consumer information and sovereignty, ability to trace and monitor GM sources for health and environmental reasons, and an endeavour to preserve the viability of different streams of agriculture, including the sanctity of organic food production. EU regulation provides for mandatory labelling, which to a large extent helps consumers to avoid consumption of GMOs, if they choose to do so. Over and above the general requirements for labelling in the EU, the Food and Feed Regulation requires labelling of foods that (1) contain or consist of GMOs, or (2) are produced from or contain ingredients produced from GMOs. Such foods are required to have a requisite label either on the package[217] or in a permanent and visible display on the shelf of sale, if it is a product not usually sold in packages.[218] It is also required to mention on the label a product's composition, nutritional values or effects, intended use or specific health concerns, when it either (1) gives rise to ethical or religious concerns, or (2) where a food is different from its conventional counterpart in characteristic or property; as and when this is specified in the authorization.[219]

In addition, the Labelling Regulation requires a 'paper trail' that traces the original transgenic process of a product to the end consumer for facilitating full, accurate and reliable labelling of such products, irrespective of the existence of living matter in the product at a later stage.[220] Such a trail is established through constitution of unique identifiers for every authorized GMO, making up a combination of (a) name of the applicant or consent holder, (b) transformation event and (c) verification

---

[217]   Article 13 (1) (a), (b) and (c) of Food and Feed Reg. It is required that (1) in cases where the product is genetically modified or produced from GM it ought to be mentioned on the cover, (2) in cases where there is more than one ingredient it should be mentioned in the list of ingredients, and (3) if there is no list of ingredients in the previous category, information that the product contains GM with particulars of the specific ingredient should be clearly stated on the cover itself.

[218]   Article 13 (1) (e) ibid.

[219]   Article 13 (2) ibid.

[220]   Article 4 (2) of the Labelling Reg.

component, expressed as nine alphanumeric digits.[221] This is relevant not only for the further facilitation of consumer information, but also helps operationalize the precautionary principle, since it assists post-release monitoring and research of hazards. It also facilitates implementation of risk management measures through post-authorization monitoring, the swift withdrawal from the market of products that may be later identified as dangerous, and also tracing the chain of production and distribution to establish causation and liability in relevant circumstances.[222] Thus any operator which places food products on the market, which (1) consist of or contain GMOs, or (2) are produced from GMOs, must transmit the specified information in writing to the operators receiving the products.[223]

While lauding the progressive nature of labelling provisions, Lee emphasizes the importance of two categories that are excluded from labelling requirements viz., products from a non-food GMO (like cotton cloth) in the traceability regulation as well as food produced from a GMO, like meat or milk from a cow fed with GM in the Food and Feed Regulation.[224] Given the actual and potential size of the market, she rightly remarks that this restricts the information flow to consumers for making meaningful, ethical and political choices. Here, it is pertinent to remember the important exemption from labelling requirements. Labelling provisions are specifically inapplicable in cases where the food or feed contains not more than 0.9 per cent transgenic material (considered for each individual GM strain) and if the producer can demonstrate that this presence is adventitious or technically unavoidable. This might be intended to be a pragmatic recognition of the fait accompli of admixture that introduction of GM crops will invariably bring. However, it poses serious questions about the integrity of the system, which claims to be designed to keep different agricultural practices separate. Discussed in the next subsection, exempting a high percentage of admixtures from labelling requirements is feared to endanger effective segregation mechanisms and the related need to secure the category of organic production of food in a meaningful manner. The 0.9 per cent threshold has been maintained even in labelling provisions of organic products under a

---

[221]   See Article 8 of the Labelling Reg., and Regulation No. 65/2004/EC of the Council establishing a System for the Development and Assignments of Unique Identifiers for Genetically Modified Organisms [2004] OJ L10/5 (Unique Identifiers Reg.). See in particular, Articles 2–6, Section A and B of the Annex.

[222]   See for instance Preamble 3 of the Labelling Reg.

[223]   See Articles 4 and 5 of the Unique Identifiers Reg.

[224]   Lee, *EU Regulation* (n. 15) 144.

different Regulation, much to the opposition of consumer groups and the Parliament, who hold that the integrity of food under the organic category may slowly erode.[225] Though these aspects appear technical and unimportant, they often constitute the heart of the matter. This becomes clearer after an examination of the coexistence guidelines that are intricately related to labelling of agricultural produce.

2.4.1 Coexistence Measures

Measures to ensure coexistence of GM, conventional and organic agriculture theoretically fall within Member State competence at present. The Directive mandates Member States to take appropriate measures to avoid unintended presence of GMOs in other products, and requires the Commission to coordinate and develop guidelines for coexistence based on studies at community and national level.[226] Given the enormous trade impact, the Directive chalks out a subsidiarity based strategy to regulate coexistence at the national level, whereby various Member States have sought to put in place appropriate measures to avoid unintended presence of GMOs in other products.[227]

Through its coordination the Commission has fettered national action on coexistence measures by placing them within the wider legal and policy framework to enforce the internal market. This is effectuated through the issue of two sets of 'non-binding' recommendations by the Commission, where it was stipulated that such measures 'should be efficient, cost effective and proportionate'.[228] Contrary to the textual

---

[225]   Regulation 834/2007/EC of the Council on Organic production and Labelling of Organic products and Repealing Regulation 2092/91/EEC [2007] OJ L189/1 (Organic Reg.). See for a detailed account of opposition: Thijs Etty, 'Current survey of substantive EC environmental law: *Biotechnology*' (2007) 7 *Yearbook of European Environmental Law* 245. See also: Maria Lee, 'The governance of coexistence between GMOs and other forms of agriculture: A purely economic issue?' (2008) 20 *Journal of Environmental Law* 198.

[226]   Article 26a Deliberate Release Dir.

[227]   See as a starting point: Thijs Etty, 'Current survey of substantive European Community law: Biotechnology' (2005) 5 *Yearbook of European Environmental Law* 293.

[228]   See Commission Recommendation 2003/556/EC on Guidelines for the Development of National Strategies and Best Practices to Ensure the Coexistence of Genetically Modified Crops with Conventional and Organic Farming [2003] OJ L 189/36, provided under A. 26a (2) of the Deliberate Release Dir., and Commission Recommendation 2010. The Court of Justice in Case C-36/11, *Pioneer Hi Bred Italia Srl* v *Ministerio dell Politiche agricole, alimentary e*

meaning of these recommendations as non-binding, the legal effect of these guidelines is counter-intuitive in nature, where the outer boundaries of Member State action are more or less hemmed in by these recommendations.[229] While most Member States are in the process of putting in place national coexistence regimes, the majority of these efforts have been earlier objected to by the Commission as potentially restricting intra-Community trade, basing its judgment on an 'extremely limited construction' of coexistence concerns.[230] Subsequent to criticism from various quarters, significantly, the Commission in 2010 emphasized the importance of protecting flexibility for Member States to take into account their regional specificities and particular needs in its recommendations.[231]

The Commission's continued insistence on coexistence measures as purely an economic issue is a significant element of the EU emphasis on segregation of crops.[232] This denies other objectives important to the public, such as pursuing precautions against ecological catastrophe

---

forestali [13], held that the recent Recommendations merely described and elaborated the earlier Recommendations, without further explanation.

[229] 'National authorities and courts are bound to take them into account, especially when they cast light on the interpretation of national measures adopted in order to implement them or where they are designed to supplement binding Community provisions': Case C-322/88 *Grimaldi* [1989] ECR 4407.

[230] European Commission, 'Commission Communication: Report on the implementation of Member States Coexistence Measures' COM (2006) 104. See Annex for detailed information on the various Member States' provisions. See further: Etty (n. 225) 377–378, who strongly criticized the Commission's interpretation of the exception under the Labelling Regulation of 'up to 0.9 per cent in-adventitious presence of GM traces' as a *minimum tolerable regulatory target* as opposed to a more correct approach of a *maximum tolerable threshold of impurity*.

[231] European Commission, 'Recommendation 2010/C 200/1 EC on Guidelines for the Development of National Coexistence measures to avoid the unintended presence of GMOs in Conventional and Organic crops' [2010] OJ C 200/1. See Recital (7): 'to better reflect the possibility ... for Member States to establish measures ... the current guidelines limit their content to the main general principles for the development of coexistence measures, recognising that Member States need sufficient flexibility to take into account their regional and national specificities and the particular local needs of conventional, organic and other types of crops'. This set of Recommendations also repealed the earlier Recommendation of 2003 (n. 228).

[232] See Annex, 2.3.2 ibid (Commission Recommendation 2010): 'Member States should consider that there may be no need to pursue specific levels of admixture where labelling a crop as GM has no economic implications'. See further: European Commission, 'Report on the Implementation of National

through keeping a form of agriculture like organic production completely independent of GMOs. The differences in perspectives on the rationale of coexistence measures are emblematic of the range of objectives that can be pursued by the coexistence, labelling and organic seed purity regimes. The effect of this insistence on coexistence as a purely economic measure is to limit the range of rationales that can inform coexistence measures. For instance, an insistence on organic as a mere economic category can thwart ethical and political aspirations of some sections of society to keep the organic form of agriculture intact as a credible public good for a number of other reasons.[233] Such reasons emanating from various quarters include public health, social and rural development, animal welfare,[234] a political choice to avoid consumption of GMOs, a form of resisting hegemonic presence by big seed companies, an ethical way of living, political activism about a non-industrialized life, and a reasonable precaution, through keeping other forms of agriculture intact for society to fall back on, against unanticipated genetic catastrophes.[235] Further, the setting of technical thresholds of contamination at 0.9 per cent could be regarded a public issue, and the same concern about thresholds is visible across the board in areas of labelling, traceability, seed purity and coexistence measures. Here, such restricted thresholds are seen by ecological groups as threatening the public credibility of preserving an important form of agriculture itself.[236] Discussions regarding how co-existence and labelling measures contribute to the pursuit of public values, and the general public's ability to contribute to the determination of these values, are engaged in the sixth chapter of this book.

2.4.2 Liability Measures

The Commission White Paper underlined the importance of civil liability regimes due to the widespread public concern about 'negative effects of

---

Measures on the Coexistence of Genetically Modified Crops with Conventional and Organic Farming' COM (2006) 104 final.

[233]   Organic Reg. (n. 225) Recital 1: 'The organic production method ... delivers a public good contributing to the protection of the environment and animal welfare, as well as to rural development'.

[234]   European Commission, 'European Action Plan for Organic Food and Farming' COM (2004) 415 final, section 1.4.

[235]   See for instance, Lee (n. 225).

[236]   See further, Les Levidow and Karin Boschert, 'Coexistence or contradiction? GM crops versus alternative agricultures in Europe' (2008) 39 *Geoforum* 174.

GMOs on our health' and the environment.[237] The Deliberate Release Directive requires its provisions to be complemented with 'liability rules for different types of environmental damage in all areas of the European Union'.[238] For activities within the purvey of EU regulation, the Commission White Paper had recommended a framework Directive providing strict liability for damage caused by certain activities categorized as dangerous activities, and fault-based liability was the preferred route for damage to biodiversity caused by non-dangerous activities.[239] However, the Environmental Liability Directive[240] has left the issue of damage caused by modern biotechnology out of its purview, except for the impact on EU designated nature conservation sites or protected species.[241] Thereby the important aspect of an appropriate liability regime for potential environmental damage is largely left to EU Member States.

The authorization procedure under the Food and Feed Regulation rules out the use of a permit defence in liability claims for GMOs, by emphasizing that 'the granting of authorization shall not lessen the general civil and criminal liability' of any food operator in respect of the food concerned.[242] In both environmental and health liability the way damage is conceptualized appears to be key; for instance, what constitutes damage to biodiversity may involve questions of mixing of GM and non-GM crops for some, since this compromises the ability of non-GM to stay GM-free. Coupled with this question of what constitutes harm, issues regarding the kind of liability (strict, fault-based or absolute) and

---

[237] European Commission, 'White Paper on Environmental Liability' 9.2.2000 COM (2000) 66 final.

[238] Recital 16 Deliberate Release Dir.

[239] Commission White Paper (n. 237). See for a useful discussion of the White Paper: Gerrit Betlem and Edward Brans, 'The future role of civil liability for environmental damage in the EU' (2002) 2 *Yearbook of European Environmental Law* 183.

[240] Directive 2004/35/EC of the European Parliament and of the Council on Environmental Liability with regard to the Prevention and Remedying of Environmental Damage [2004] OJ L 143/56 (Liability Dir.).

[241] Christopher Rodgers, 'Implementing the Community Environmental Liability Directive: GMOs and the problem of unknown risk' in Bodiguel and Cardwell (n. 28) 198. See further: Thijs Etty, 'Current survey of substantive European Community law: Biotechnology' (2006) 6 *Yearbook of European Environmental Law* 252. He discusses the slim chance of the Commission including possible damages caused by GMOs in the application of the Directive, through administrative provisions in 2014, *qua*, Art. 18 (2) and 3 (b) of Liability Dir.

[242] Articles 7 (7) and 19 (7).

what kind of compensation ought to be appropriate, continue to be amorphous at this stage, and are largely unclear even in Member States' legal regimes.[243] The extent and limits of the possibilities of a more general public having a say in these issues of liability are, naturally, dependent on the respective Member State legal regimes.

2.4.3   An 'Opt-out' from GM Agriculture?

A recent development of significance to public involvement and participation is the Commission's proposal to introduce additional safeguard measures in the Directive, providing for Member States to restrict or prohibit the cultivation of GMOs that are already authorized by EU regulation.[244] The proposal emerged in the context of the continuation of unauthorized safeguard bans by a number of Member States,[245] and is subsequent to suggestions by thirteen Member States to the Environment Council for introduction of legislation to opt out of cultivation of specific GMOs.[246] Further, the Kingdom of Netherlands made declarations to the Agriculture and Environmental Councils, asking the Commission to suggest a solution on cultivation while taking into account the socioeconomic dimensions of GMO cultivation.[247] Proposed in 2010, the clause would provide a new safeguard clause under the Directive for Member States to restrict or prohibit cultivation of GM crops in their territories on specific grounds outside of science, given the strong local and regional dimension of cultivation.[248] Importantly, it bars the consideration of protection of health and the environment under the new proposal,

---

[243]   Etty (n. 225) 2007.

[244]   See Communication from the Commission to the European Parliament, the Council, the economic and social committee and the Committee of the regions on the freedom for Member States to decide on the cultivation of genetically modified crops, COM (2010) 375 final.

[245]   See text near nn. 181–185.

[246]   Vide Council meetings of 2 March, 23 March and 25 June 2009, note with ref. 11226/2/09 REV 2 of the EU Council, cited in Commission Communication (n. 244) fn 12.

[247]   Note with ref. 7581/09 of the EU Council, cited in Commission Communication (n. 244) fn 10.

[248]   European Commission, 'Proposal for a Regulation of the European Parliament and of the Council amending Directive 2001/18/EC as regards the possibility for the Member States to restrict or prohibit the cultivation of GMOs in their territory' COM (2010) 375 final, Article 1 and Recital 8.

and explicitly excludes the application of these safeguards to the commercial use of authorized GM seeds for purposes other than for cultivation.[249]

Since the balance may not always be right between an EU framework and the need to take account of diversity of cultivation in twenty-seven Member States, the purported aim is to combine 'science based EU authorisation' with freedom for Member States to 'decide whether or not they *wish* to cultivate GM crops on their territory'.[250] The proposal pointed out that considerations about cultivation can only have an impact on the way authorizations are adopted at EU level at present, notwithstanding that the existing EU framework 'could give margin to consideration of specific conditions of cultivation in Member States', viz., 'consideration of regional aspects under the risk assessment and conditions of authorisations, or of other legitimate factors under the Regulation [sic]'.[251] The Commission opined that these existing considerations may be too restrictive in fulfilling 'the central notion' of allowing Member States some flexibility and margin of appreciation to decide on cultivation of authorized GMOs, 'taking into consideration their specific conditions'.[252]

Notably, the proposal stipulates that the proposed safeguard measures shall not be based on grounds covered by the environmental risk assessment under the original EU authorization, or on grounds related to avoiding the unintended presence of GMOs in other products.[253] It further stipulates that the safeguard measures shall respect the Treaties, including the principle of non-discrimination under Article 34 and 36 TFEU,[254] as well as relevant international obligations, in particular with those established under the WTO.[255] Though the acceptable contours of conformity and deviance from these international obligations will be examined in the fifth chapter, it is significant that the proposal expressly reserved the assessment of potential health and environmental risks to

---

[249] Explanatory memorandum, Commission Communication (n. 244) 5, 2.2.1.B.

[250] Ibid 2.

[251] Ibid 3, 2.1.A.

[252] Ibid 3, 2.1.A.

[253] Ibid 6, 3.1.

[254] See Maria Lee, *EU Environmental Law, Governance and Decision Making* (forthcoming, Hart 2014), Chapter 10, text near nn. 93–103.

[255] Explanatory memorandum (n. 244) 7, 3.1.2.

EFSA, and away from the ambit of this safeguard clause. This over-insistence on compliance with relevant international and Treaty obliga-tions is sometimes taken as a cue to articulate concerns that this proposal is either 'essentially pointless or even a cynical attempt to hasten authorization, in confident expectation that any autonomy will be mean-ingless'.[256] While this may or may not be the case, the reference to public values like social, economic and ethical aspects in decision-making on GM agriculture, and that too with an explicit emphasis on public involvement, is profound.

The document seeks a right balance between maintaining the EU system of authorizations based on the scientific assessment of health and environmental risks, and the need to grant freedom to Member States to address specific national or local aspects raised by the cultivation of GMOs.[257] Which aspects can be considered as grounds for safeguard measures under the proposal in relation to the 'Member [States'] *wish* to cultivate GM crops on their territory' needs further clarity.[258] The explanatory memorandum delineates 'social, economic and ethical aspects to … provide the platform for respective decisions'.[259] Social, economic and ethical aspects are expected to be put on the table and provide the platform for the respective decisions at national, regional or local level.[260] While the use of social or ethical values as a basis to restrict GM cultivation would need further elaboration, the theoretical possibility for a legal ban exists. This opening is implicit in the Court's judgment in *Commission* v *Poland.*[261] In this decision, though the Court rejected Poland's reliance on ethical and religious requirement as a justification of its ban on GM seeds, the rejection was only on the grounds that Poland failed to demonstrate that these requirements were the real basis for its regulatory measures. Promisingly, the Commission expects an increase in the level of public involvement in national and regional decision-making through this clause in the ascertainment of these social, economic and ethical aspects. The Commission emphasizes its expectation that the Member States will allocate more resources and time to involve their publics with regard to these decisions.

---

[256] Lee, 'Governance and Decision Making' (n. 254), Chapter 10, text near fn 118.
[257] Explanatory memorandum (n. 249) 7, 2.3.
[258] Explanatory memorandum (n. 249) 7, 2.2.2, emphasis supplied.
[259] Ibid.
[260] Ibid.
[261] *Commission* v *Poland* (n. 190).

To sum up, the general effort in this chapter has been to provide a cursory introduction to the EU regulatory framework on GM food and feed, with an emphasis on identifying room for public participation. The chapter noted the particular visibility of the regulatory objective of safety in public decisions about use of GMOs. Here, the regulatory tool of risk dominates the way the objective of safety is pursued. This is discernible during both authorizations for release of GMOs and the subsequent environmental guarantee and product safeguard measures. There is also an emphasis on the recognition of the precautionary principle in this regulatory pursuit of safety. Apart from this, labelling and coexistence measures were identified as having the potential to represent and pursue other public values in regulation. How wider publics are allowed to participate in the identification and representation of other values is further examined in the sixth chapter. Coming back, the Deliberate Release Directive and the Food and Feed Regulation have provided rights for the general public to be informed and consulted during the authorization procedure,[262] as well as in some circumstances for Contained Use[263] and field trials.[264] These are the only spaces where issues with classical representation of traditional expert-based safety regulation are acknowledged, through allusions and recourse to public access to information and consultation in decisions related to release of GMOs. Legal opportunities for public participation are important arenas for the expression of public contestations and disagreements about development and use of GMOs in agriculture and food. Whether such information and consultation can amount to meaningful participation and representation during the regulatory employment of the tool of risk needs further reflection. Towards this end, it is imperative that we seek to understand the regulatory tool of risk by conceptually navigating the employment of this concept. This is attempted in the next chapter. Subsequent to it, the fourth chapter builds on this discussion by focusing on the use of the precautionary principle in environmental law, and how it has been employed in EU regulation of GMOs.

---

[262] Text to n. 150, and nn. 156–159.
[263] Text to n. 175.
[264] Text to nn. 179–180.

3. Risk, science and society

> Risk theory needs to be interested in investigating the forms of knowledge, dominant discourses and expert techniques and institutions that serve to render risk calculable and knowable.[265]

Protection of human health and the environment against hazards from the introduction of new technologies has long been identified as a central function of the modern welfare state.[266] The resolution of safety concerns had become particularly important for the acceptability and feasibility of the development and application of new technologies by the 1980s. Since awareness of industrial and technological hazards had grown significantly by this time, many treated the concerns about introduction of new technologies like biotechnologies and nano-technologies as restricted only to an issue of safety.[267] Significantly, the last three decades have seen a slow and steady change in regulatory emphasis on safety, from an investigation of hazards to a more complex regulation of risks,[268] where risk is seen as 'replacing need as the core principle of social policy formulation'.[269] The introductory account of EU regulation of GMOs in the previous chapter identified the pursuit of safety as the dominant trope for authorization and release of GMOs. Two motifs were further discernible, viz., the concept of risk and the principle of precaution, as central to the way concerns of safety are represented and mediated in EU regulation. This chapter seeks a conceptual focus on risk as a dominant regulatory tool to pursue the public objective of safety. The endeavour

---

[265] Deborah Lupton, *Risk and Socio-cultural Theory* (Cambridge University Press 1999) 6.

[266] See for instance: Hazel Kemshall, *Risk, Social Policy and Welfare* (Open University Press 2002) chs 1 and 2.

[267] Chauncey Starr, 'Social benefit versus technological risk: What is our society willing to pay for safety?' (1969) 165 *Science* 1232.

[268] Here hazard is understood as the intrinsic potential to cause an identified adverse effect whereas risk is understood, in a crude sense, as the likelihood of its occurrence. See for a useful treatment of this shift: Lakshman Guruswamy, 'Sustainable agriculture: Do GMOs imperil biosafety?' (2002) 9 *Indiana Journal of Global Legal Studies* 461, 484.

[269] Kemshall (n. 266) 10.

here is to understand how risk is employed in regulatory engagements with public contestations about GMOs, and the technological trajectories GMOs signify.

Evidenced in our examination of the EU regulatory framework in the previous chapter, risk is currently the central organizing principle in the commercial release of GMOs. Risk regulation appears to trump all concerns of safety in the release of GMOs, where an ascription of safety through risk regulation is deemed as a sufficient condition for the commercialization of a new GMO. The Deliberate Release Directive and the Food and Feed Regulation provide for bifurcated risk regulation (by means of technical analysis and political management) through a maze of institutional procedures, to ascertain the acceptability of environmental and health risks of a decision to permit commercialization. The aspiration in this bifurcation is that a prior technical characterization of the risks involved is subsequently implemented through political management, which will also take into account public preferences and concerns. The current chapter is a forensic assay through prominent conceptualizations of risk, including the Risk Society thesis, techno-scientific frames of risk and the Governmentality thesis, to argue that public participation lies at the conceptual core of risk. Further, the norms of trust and legitimacy are contended to be necessary for the establishment of an ascription of risk.

An important confusion involving the term 'risk' lies in it being used to denote both a state of uncertainty and a process of identifying danger:

> Theorists using the term risk are divided in whether the most significant thing about this term is thought to be threat or uncertainty which it implies, or else (on the contrary) the means of responding to threats and uncertainties which its use brings with it ... Consequently, they are divided on the question whether risk encapsulates the weakness of human decision-makers in the face of hazards and the unknown, thus compromising the power to act; or whether it encapsulates the very opposite, the route by which human beings may achieve control over their own destinies, providing the means to action and decision.[270]

By identifying this confusion and discussing the concept through the aforementioned sociological approaches to risk, the argument of the necessity of public participation in the construction of risk is furthered in the current chapter. First, the chapter identifies important assumptions of techno-scientific approaches to risk, certain related fundamental problems, and the difficulties such understandings pose to possibilities of wider representation and participation in public decision-making about

---

[270]   Jenny Steele, *Risks and Legal Theory* (Hart 2004) 3.

new technologies like GMOs. Here, the assumptions behind bifurcation of risk and deficit models are examined, and the fundamental weaknesses of the assumptions of the classical risk paradigm identified. The point here is not merely to observe that techno-scientific ascriptions of risk are conceptually erroneous, and often even unscientific, and hence that policy needs to take into account these deficiencies to make it more scientific. In addition, if one seeks a keen understanding of the use of the tool of risk in the regulation of new technologies, we need to move beyond these inadequate conceptualizations of risk, into examining what are broadly categorized as sociological conceptualizations of risk. Once risk is seen as an estimation that depends on a number of assumptions about uncertain futures, as elaborated later, contestations about risk need not be about public ignorance and lack of awareness of the facts, but could be about a conflict of values, norms and priorities.

Secondly, important sociological approaches are explored to gain a keener understanding of how the concept of risk is employed, assuming that GMOs are part of the aforementioned risk contestations about values, norms and priorities. Through this exploration, the norms of trust and engagement are identified as central to ascriptions of safety for them to be effective. In other words, an emphasis on trust and engagement is not just to make a pre-existing ascription of risk work, but rather, they are necessary to construct a workable ascription of risk. This is in contrast to the use of public participation for better management and implementation of prior constructions of risk by technical bodies. Over and above an attempt to bolster the legitimacy of technical risk ascriptions through public engagement exercises, it is asserted in this chapter that an appropriate understanding of the tool of risk necessitates public trust and engagement in the core of its construction itself. Towards this end, the chapter engages with ideas of Luhmann, Beck, Douglas and Governmentality scholars in three subsequent sections of this chapter.

3.1   TECHNO-SCIENTIFIC APPROACHES TO RISK

Conventionally, most descriptions of risk start with techno-scientific approaches, and very often end with them as well. Such technocratic approaches emerged from disciplines such as science, engineering, psychology, economics, medicine and epidemiology. In this realist approach to risk, risk is largely assumed as an objective phenomenon that can be accessed through expert investigation and assessment, as well as a tool to measure and respond to uncertainty. The focus of research on risk in these fields is the identification of risk, mapping their causal factors, building predictive models of risk relation and people's responses to various types of risk, and proposing ways of limiting the effects of risk.[271] Risk here is seen as the probability that a particular adverse event (identified as a hazard) occurs during a stated period of time; 'as a probability in the sense of statistical theory, risk obeys all formal laws of combining probabilities'.[272] This is a *reductive-aggregative* process as it involves 'simplification of multivalent complexities to simple parameters of likelihood and magnitude, and subsequent aggregation across highly diverse dimensions, contexts and etiologies' to reach an apparently transcendent quantitative idiom, which is then taken as synonymous with objective authority.[273]

Techno-scientific approaches to risk assume expert measurement and calculation as the most appropriate starting point,[274] where risk assessment is understood to be a series of steps including hazard identification, hazard characterization, exposure assessment and risk characterization. Here, Knight's differentiation of risk and uncertainty has continued to be an important influence, viz., if you do not know for sure what will happen, but you know the odds, that's risk, but if you do not even know the odds, that's uncertainty.[275] There is a consensus among this kind of risk literature about the objective nature of the thing, and the perception of the thing is independent of the object itself. To the contrary, Adams reminds us that despite the formal requirement that the odds be known so that numbers are attachable to the probabilities and magnitudes of possible outcomes of risks and uncertainty in game theory, operations research, economics or management science, they are virtually always assumed or invented in practice since such numbers are rarely available.[276]

---

[271] Lupton, (n. 265) 2.

[272] Royal Society (Great Britain), *Risk Assessment: Report of a Royal Society Study Group* (Royal Society 1983). The emphasis on statistical probability continues among assessment by techno-scientific institutions: see for instance, EFSA, 'Guidance on the environmental risk assessment of genetically modified plants', <http://www.efsa.europa.eu/en/scdocs/scdoc/1879.htm> last accessed 21 August 2011.

[273] Stirling (n. 60) 96.

[274] Lupton (n. 265) 2; John Tulloch, 'Culture and risk' in Jens O. Zinn (ed.), *Social Theories of Risk and Uncertainty* (Blackwell 2008) 138, 141.

[275] Frank Knight, *Risk, Uncertainty and Profit* (Hart, Schaffner and Marx essays, Houghton Mifflin Co. 1921) ch. VII.

[276] John Adams, *Risk* (UCL Press 1995) 25–26, also ch. 6 in general. Further, 'early theorists drew a distinction between risky events where probabilities are known by science, and uncertain events where the probabilities of events are unknown. This distinction has been eroded owing to the emergence of subjective

3.1.1   A Bundle of Incertitudes

The central problem with this conception of risk, viz., finding the quotient of the probability of the occurrence of an identified hazard, is the incomplete and unstable nature of knowledge that underlies both the fundamental parameters of hazards and likelihood. The effect of the instability of underlying knowledge on a full characterization of risk is seen to be the loss of any meaningful scientific rigour at a fundamental level. Stirling describes the diversity in the circumstances of knowledge of these parameters that are conventionally rolled together under the term 'risk'.[277] He terms the cluster of these contrasting states of incomplete knowledge 'incertitudes', and identifies the strict techno-scientific definition of risk (in other words, of a probabilitization of identifiable hazards) as only one among them. According to Stirling, the incertitude of risk is possibly applicable in many important fields where relevant past experience or reliability of scientific models foster high confidence, and where conventional reductive aggregative techniques like risk assessment actually use a scientifically rigorous approach. However, in the other three states of incertitude, namely uncertainty, ambiguity and ignorance, Stirling demonstrates that the traditional employment of risk is simply not applicable, and is unscientific.

Stirling's description of uncertainty is a situation in which there is confidence in the characterization of the different possibilities of outcome, but the available empirical information or analytical models do not present a definitive basis for assigning probabilities. Though probability cannot be assigned in these circumstances, it is very possible that regulatory bodies may choose one among many equally plausible outcomes to arrive at a regulatory decision. This selection of one among various possibilities by the regulator can be through the exercise of their political or ethical judgement. Here Stirling comments that the scientifically rigorous approach would be to acknowledge the open nature of a variety of different possible interpretations and the exercise of subjective judgement. This acknowledgement is in contrast to a non-transparent

---

(judgmental) notions of probability ... yet many analysts continue to distinguish different types of uncertainty in their analysis, such as not knowing the particular draw from a known probability distribution from which draws will be taken. The latter type of uncertainty is particularly common in risk assessment': John D. Graham, Per-Olov Johansson and Junko Nakanish, 'The role of efficiency in risk management' in Timothy McDaniel and Mitchell J. Small (eds), *Risk Analysis and Society* (Cambridge University Press 2004) 258–259.

[277]   Stirling (n. 60) 98.

reduction of these various possibilities to a single one, as if there is a single expected value or prescriptive recommendation that is incontrovertible in situations of uncertainty. The incertitude of ambiguity is an ideal type that denotes a condition where it is the characterization of outcomes that is problematic, as opposed to the characterization of probabilities among uncertainties. To ensure a scientifically rigorous process here, the disagreements over the selection, partition, bounding, measurement, prioritization or interpretation of outcomes require attention to be extended to 'broader values, epistemologies and ontologies'.[278] Stirling cites the contending ways of framing issues in GM food regulation, viz., the differential invocation of ecological, agronomic, safety, economic, social or even more fundamental values concerning relationships and responsibilities within society, nature and so on, as a very good example of an instance of the operation of the incertitude of ambiguity. The final condition of incertitude in this matrix – ignorance – is a state in which there is an inability to fully characterize both probabilities and outcomes. Ignorance is differentiated from uncertainty in that in cases of uncertainty the focus is on agreed known parameters like say, carcinogenicity, as opposed to unknown or unknowable outcomes in the former case. Ignorance is sought to be distinguished from ambiguity in that the parameters are not just contestable or indeterminate, as in a situation of ambiguity, but in contrast are unbound and at least partially unknown. Stirling remarks that some of the most important contemporary environmental issues circle around ignorance, like in the early histories of ozone depletion and endocrine disrupting chemicals.[279]

Currently under the techno-scientific paradigm, in the name of risk analysis, all four aforementioned ideal types of incertitude are undertaken under the single category of risk. The point this raises is not merely that this practice is not even minimally scientifically rigorous, but as regards the more important and powerful implications of the invocation of science in risk regulation. Given the paucity of directly applicable data in radically new technologies, like, say, biotechnology or synthetic biology, where the data used are usually incomplete, absent or even beyond our present understanding, it is only through a denial of the unstable nature of the underlying knowledge and data that a supposedly unitary and quantified techno-scientific artefact of risk can be arrived at. The regulatory tool of risk analysis in this techno-scientific paradigm is a convenient alibi for political managers from the considerable conceptual

---

[278] Ibid 99–100.
[279] Ibid.

difficulties encountered during the regulation of safety of radically new technologies. In some cases it could also assuage public fears, with or without scientific basis. Since such incertitudes are more or less inevitable in the introduction of radically new technologies today, the supposed technical assessment of risks has to rise to the realm of values and norms, and not mere measurement and calculation.

This recognition of the normativity of the risk assessment is in sharp contrast to the strict divide between scientific risk assessment (fact) and political risk management (value) prevalent in most regulatory frameworks today, and central also to the risk regulation of GMOs in Europe. Thus, despite the well-made argument that technical risk assessment is permeated with normative assumptions,[280] the techno-scientific conceptualization of risk has continued to be the dominant regulatory tool in avoiding harm to the environment and human health. The representational quality of such a conceptualization of risk is under serious strain, when one recognizes the normative choices in risk assessment and the systemic incertitudes inherent in the development of such technologies.

3.1.2 Bifurcation and Deficit Models

Important facets of this techno-scientific conceptualization include the bifurcation of risk regulation into scientific assessment and political management, and the deficit models that often accompany them as a solution to problems that arise from bifurcation.[281] This bifurcation of risk regulation famously gained currency in the Red Book of the US National Research Council. This report characterized the potential adverse health effects of human exposure to toxic substances, where technical risk assessment was partitioned from risk management: risk assessment culminates in quantitatively or qualitatively estimating the incidence and magnitude of an adverse effect on a given population, whereas risk management, drawing upon risk assessment, culminates in developing policy options and evaluating their health, economic, social and political implications.[282] The 'clear conceptual difference between

---

[280] Carl Cranor, 'The normative nature of risk assessment' (1997) 8 *Risk, Health, Safety and Environment* 123. See further, Expert Group on Science and Governance (n. 53) ch 3.

[281] See for an introduction on deficit models, text to n. 46.

[282] National Research Council (US), *Risk Assessment in the Federal Government: Managing the Process* (Committee on the Institutional Means for Assessment of Risks to Public Health, National Academy Press 1983).

assessment of risks and the consideration of risk management alternatives' sought to be maintained was so stark that even a perception that the latter influenced the former was supposed to lead to a serious lack of credibility.[283] The British Royal Society report, in the same year and in a similar vein, defined risk as a detriment that is the compound of probability and magnitude of an adverse event.[284]

This institutionalization of the fact–value divide in the risk architecture set out in the Red Book is often identified as the 'principal point of reference' for regulatory authorities, despite subsequent shifts of nuance and emphasis.[285] Adams points to the continuation of this distinction in various important British and American official reports, between 'the sort of thing the *experts* know about' and 'the *lay* person's often very different anticipation of future events'. For instance, he discusses the subsequent calls for bridging the gap between what is stated to be scientific and the way in which the public gauges and makes decisions, in the terms of reference of the 1992 UK Royal Society report. He refers to the failure of the Committee in this, where the first four chapters continue the aforesaid distinction, while curiously in Chapter 5 this approach is flatly rejected: 'that the view that a separation can be maintained between objective risk and subjective or perceived risk has come under increasing attack, to the extent that it is no longer a mainstream position'.[286] In this chapter of the report, risk is understood as culturally constructed, in which both the adverse nature of particular events identified as hazards, and their probability are inherently subjective. The assumptions in the bifurcation belie the inevitable normative commitments behind technical risk assessment that may take the form of 'subjective judgments, influential social values, contestable assumptions and administrative procedures that are open to contingent framings and the tacit or deliberate exercise of power', where political interests as well as habitual ways of institutional and disciplinary thinking can shape risk assessment in determinative ways.[287] Despite such possible shifts in understanding, bifurcation continues to be the backbone of GMO regulation in Europe till date, and the

---

[283] Adams (n. 276) 27.

[284] Royal Society (n. 272) 260.

[285] Expert Group on Science and Governance (n. 53) 32.

[286] Adams (n. 276) 8–9.

[287] See for a detailed critique of the bifurcation: Deborah G. Mayo and Rachelle D. Hollander (eds), *Acceptable Evidence: Science and Values in Risk Assessment* (Oxford University Press 1991); Ellen Silbergeld, 'The uses and abuses of scientific uncertainty in risk assessment' (1986) 2 *Natural Resources and Environment* 17.

default regulatory architecture of institutions governing environmental and health risks in most parts of the world. Notwithstanding the fact that in more recent official reports, the line between the two is drawn less clearly, where it is often argued that an understanding of both is needed for a full characterization of risk,[288] the assumptions informing this bifurcation have continued to shape regulatory regimes in Europe and the US. This similarity is despite descriptions that Europe and the US have diametrically opposite risk regimes, where one is claimed to be precautionary, the other is supposed to be science-based.[289]

Within this bifurcated regulatory framework, public concerns about safety in most parts of Europe appear not to be assuaged by repeated positive reports of risk assessment from scientific authorities like EFSA.[290] Given the backdrop of the institutionalization of the fact–value divide, this situation is sought to be remedied in policy circles, by either attempting to deal with normative aspects in the risk management (since it is after all the political stage), or through improvements in risk assessment procedures to explicitly account for social, ethical and political concerns. This has elicited three kinds of suggestions: (1) risk communication efforts, that seek to educate and inform the lay public about the real nature of the risks, identified through the technical risk assessment, as opposed to their current perceptions, branded as either false, irrational or ill informed; (2) explore possibilities to include lay perceptions in the risk assessment to make it more acceptable to the public; and (3) make risk assessment more scientific, say by acknowledging the real nature of instabilities of base data, possible ambiguities and ignorance, so as to make this more transparent and acceptable to the critics and opponents of the process. All three attempts to fill the deficits in communication or scientific rigour fall within a broad techno-scientific understanding of risk that is placed in a bifurcated institutional arrangement.

While the project of the techno-scientific conceptualization is to describe the possibilities of danger in a statistical way, what is decided to

---

[288] National Research Council (US), *Understanding Risk* (Committee on Risk Characterization, Paul C. Stern and Harvey V. Fineberg eds, National Academy Press 1996).

[289] See for instance, David Vogel, 'Risk regulation in Europe and the United States' (2003) 3 *Yearbook of European Environmental Law* 1. See further, Andrew Stirling and Patrick van Zwanenberg, 'Risk regulation in Europe and the United States – A response to David Vogel' (2003) 3 *Yearbook of European Environmental Law* 47.

[290] Text to nn. 161–163; nn. 28, 53, 54.

be chosen or ignored as dangerous, and the related base line of such selection, is normative, political and cultural at once. Even when there is a recognition that normative issues are invariably interlaced with technical issues that render the former invisible, the prospect of studying the technical details of, say, toxicology, diminishes the eagerness of policymakers to raise such normative issues.[291] Despite its apparent convenience in the short term, however, such avoidance is democratically inappropriate, conceptually incoherent and practically ineffective to represent public contestations about technology. The point here is not merely to observe that techno-scientific ascriptions of risk are conceptually erroneous and often unscientific,[292] and hence that policy needs to take into account these deficiencies to make it more scientific. In addition, if one seeks a more appropriate representation during law's use of the tool of risk, and possible room for public participation in decision-making on GMOs, we need to move beyond these inadequate conceptualizations of risk to examine what can be crudely clubbed as sociological conceptualizations of risk.[293]

3.2 ENGAGEMENTS WITH SOCIOLOGICAL APPROACHES TO RISK

The next few sections seek to reach a pertinent understanding of how legal systems employ the tool of risk, and the accompanying issues of representation, by forensically identifying conceptualizations of risk that are loosely termed 'sociological approaches'. The attempt here is a forensic examination of influential sociological understandings of risk within the context of the public disquiet about GMOs in the EU. Such a forensic examination is sought towards an investigation of the nature of the exclusions and legitimizations possible through risk regulation. In particular the centrality of public trust, participation and engagement

---

[291]   WRR (n. 2) 80.

[292]   These techniques as a pursuit of an objective artefact that can be discovered through expert techniques are unscientific in the case of radically new technologies, due to the unstable and incomplete nature of the base-data upon which they have necessarily to base themselves. Deficit models built on this myth about the Holy Grail of expert risk, which needs to be communicated to the public for their education, are therefore fallacious.

[293]   Such an exploration becomes even more urgent once we recognize post-normality of science, which is described in Chapter 1, as a valid sociological assumption of the contemporary.

during the use of the tool of risk in GMO regulation is sought to be emphasized through this forensic exploration.

Conflicts about production of risk and production of wealth may be intricately interwoven, where creation of wealth could inevitably lead to concerns of risk, and conversely identification and production of risk have distributional issues, with multifarious connections to wealth. For instance, starting an industrial plant may create additional wealth for a society, but it can also create hazardous conditions to the neighbouring communities. How this additional wealth is allocated in a society may lead to social conflicts, and so might how the related hazards are distributed in the society – be it a territorial distribution or a class-based distribution. Whether conflicts regarding the two are connected depends on the social context, and how law represents them.[294] Such distributional and related questions of justice that accompany the essentially anthropogenic nature of the source of technological risks are another important reason that may compel us to engage with sociological and constructivist approaches to risk.

Adopting a cultural anthropological approach, Mary Douglas set forth an influential perspective on risk that introduced the crucial importance of cultural foundations into the risk discourse, a discourse which until the 1970s was dominated by technical-scientific and cognitive-rational approaches to risk regulation.[295] By posing an important question as to why certain risks are emphasized, putting particular and significant emphasis on one possibility over another, Douglas connects inquisitions about risk to forms of social organization that are either defended or attacked. Often termed 'foundational' within internal histories of risk theory, her concern did not concentrate much on the reality of risk (in other words, if and how, it exists as concrete entities) but how it is politicized.[296]

---

[294]   See further: Mary R. English, 'Environmental risk and justice' in Timothy McDaniels and Mitchell J. Small (eds), *Risk Analysis and Society: An Interdisciplinary Characterisation of the Field* (Cambridge University Press 2004) 119.

[295]   Mary Douglas and Aaron Wildavsky, *Risk and Culture* (University of California Press 1982); Mary Douglas, *Risk Acceptability According to the Social Sciences* (Routledge and Kegan Paul 1985); Mary Douglas, 'Risk as a forensic resource' (1990) 119 *Daedalus* 1.

[296]   She asserted that risk is theoretically independent of objective reality and has no immediate impact on the forms of critique and resistance to technology, by focusing on how certain risk inquisitions are politically preferred, as opposed to other inquisitions that are equally plausible. Tulloch reminds us, however, that there also exists a strong realist streak in her work, where not all is culture,

3.2.1 A Matter of Attribution

The German sociologist, Niklas Luhmann in some ways built on this powerful critique about conceptualization of risk among techno-scientific approaches, which Douglas had initiated through her assertion of the context of politics and culture in the reification of risk, despite considerable differences in their respective projects including variations in the epistemic object of risk within these projects. Since Luhmann's systems theory pays attention to 'the constitutive character of risk for modern society', naturally, he does not conceptualize risk as an objectively given danger.[297] In his conceptual separation between risk and danger, the distinction between the two is not objectively given, and further is predicated on a first order differentiation of risk and safety. Safety as a material state of being (like health) is a goal that is not really achievable, and is only understandable in relation to its opposite, uncertainty (like illness). Here, safety is described as a state which is also never certain, wherein by choosing the apparently safe option one cannot be sure whether another option may not have provided more advantageous or even safer opportunities.

This uncertainty of both safety and risk, coupled with another concern that such uncertainty does not freeze decision-making at large, makes a second order distinction of risk and danger necessary for Luhmann. Illustrations of decisions related to installation of a nuclear power plant and solemnization of a marriage are used to show how uncertainty exists in an everyday sense, which requires that decisions have to be taken in spite of uncertainty. For decisions to be made even with uncertain and incomplete information, entailing uncertain consequences as in the future of a marriage, he reasons, we are prepared to take a decision without being absolutely sure only when we are in control. Thus for those in control of a decision, the consequences of which they cannot be absolutely sure, risk implies the uncertainty of the consequences of their decisions. On the contrary, negative effects caused by external events over which one has no control, for instance of a community living next to a nuclear plant that may not have any say in the institution or running of the plant, the uncertainty and possible negative effects may be regarded

---

situating different cultural responses to risk within a structuralist grid group model of social organization that is optimistic of the resilience of the institutional structures against the cultural emphasis of global doom narratives (n. 274) 142–143.

[297]   Klaus P. Japp and Isabel Kusche, 'Systems theory and risk' in Jens O. Zinn (ed.), *Social Theories of Risk and Uncertainty* (Blackwell 2008) 76, 80.

as dangers. The difference between risk and danger, here, is not then about uncertainty but about attribution as to whether one had control over a decision that subsequently led to an undesirable event. Thus risks taken by some can become dangerous to others due to potential harm caused to others who do not identify themselves with the decision (and thus the attribution of risk), while those who identify themselves with the decision (thus those who feel themselves to be in control) would attribute the potential harm of the decision as risk.[298] In cases where the damage is not voluntarily accepted, the distinction between decision-makers who identify something specific as risk, and those who feel affected by it, becomes stark.[299]

The decision-maker may evaluate and take a risk, which may become a danger for someone, either to her personhood or something which she values. Thus it is the decision-maker who would appear to have caused the potential harm, though who the decision-maker is, again depends on the attribution.[300] In an instance of genetic modification of crops, is the biotech company the decision-maker for developing a crop technology in a specific technological pathway, as opposed to, say, a farmer who ought to have used the technology in a specific manner or the regulatory agencies that evaluated the release application? Ralph Nader's efforts to challenge the long-term supposition regarding motor vehicle accidents in the US that it is the reckless driver who is the dominant risk object/ decision-maker is a good illustration to make this point. He pointed out several alternative possibilities of risk objects, including the car manufacturers that insist on having particular safety and operational designs, and the road planners and traffic designers who might make attributions of risk that might be dangerous for users.[301] Here, the risk–danger distinction made by Luhmann coincides with differences in political and normative priorities between those making a decision and those affected by it. This can complicate the role law plays in the regulation of technologies where, devoid of popular participation, a simple claim to procedural democracy based on expert advice may be insufficient to address these normative questions.[302]

---

[298]   Ibid 88.

[299]   Niklas Luhmann, *Risk: A Sociological Theory* (A. de Gruyter 1993) 104.

[300]   Japp and Kusche (n. 297) 87.

[301]   Stephen Hilgartner, 'The social construction of risk objects: or, how to pry open networks of risk' in James F. Short and Lee Clarke (eds), *Organisations, Uncertainties, and Risk* (Westview Press 1992) 39.

[302]   See for more on this aspect: Ulrich Beck, 'Environment, knowledge and indeterminacy: Beyond modernist ecology' in Scott Lash, Bronislaw Szerszynski

Returning to the issue of arriving at an appropriate manner to understand how risk is employed in the EU regulatory framework for release of GMOs, it was earlier noted that despite repeated positive risk assessment reports from EFSA, considerable sections of the public have repeatedly feared the releases to be unsafe. In this regard, both Douglas' and Luhmann's insights can take us forward in understanding this impasse in representation of regulatory concerns. The cultural turn spurred by Douglas' work, as well as the second order differentiation of risk and danger and the emphasis of risk as attribution by Luhmann, help us understand the employment of the concept of risk in legal systems in a more pertinent way for contestations and controversies about technological safety than the techno-scientific conceptualizations.

Amid the competing characterizations of risk following Douglas, two streams of scholarship that are relevant for our endeavour have recently gained currency. First is Governmentality scholarship, which explores ways in which the state and other governmental apparatuses work together to manage and regulate populations via discourses and strategies. They have delineated risk as part of a specific kind of liberal governance, also suggesting a broader theoretical approach to management of uncertainty beyond risk.[303] Secondly the Risk Society thesis, primarily ascribed to Ulrich Beck and to a lesser extent to Anthony Giddens and various others, has provided a major impetus to recent sociological examination of risk, especially risks arising from new technologies.

3.3 TECHNO-SCIENCE AND THE RISK SOCIETY THESIS

It is apposite that a peek into the Risk Society thesis in the context of GMO regulation is attempted through the reflections of the German sociologist Ulrich Beck, given his particular emphasis on technological

---

and Brian Wynne (eds), *Risk, Environment and Modernity: Towards a New Ecology* (Sage 1996) 37.

[303] See for a useful introduction: Mitchell Dean, 'Sociology after society' in David Owen (ed.), *Sociology after Post-modernism* (Sage 1997) 205. The tradition of risk scholarship around Luhmannian systems theory, discussed in the previous section, has inevitable connections to the governmentality project, 'given the questions that are asked by Luhmann and Japp about modern forms of observing decisions about safety'. See Jens O. Zinn, 'Introduction' in Jens O. Zinn (ed.), *Social Theories of Risk and Uncertainty* (Blackwell 2008) 1, 15.

risks, and the breadth of his contribution.[304] Along with his limited collaboration with the English sociologist Anthony Giddens in the 1990s, these contributions have had considerable influence on imaginations of risk regulation in contemporary times. His identification of prominent inadequacies of risk governance is an important starting point for a more appropriate representation and mediation of anxieties regarding introduction of new technologies. These facets include a focus on erosion of the monopoly of truth claims of techno-scientific knowledge for practical application, growing transnational political regulation in issues of technological hazards, an acknowledgement of the relevance of different forms of everyday knowledge outside of organized science and an examination of the shifting boundaries of our understandings on nature and society.

An important part of Beck's work focuses on new technological risks, with which he leads us towards his general thesis on contemporary modes of general social change.[305] His core thesis is that:

> In advanced modernity the social production of wealth is systematically accompanied by the social production of risk. Accordingly, the problems and conflicts relating to distribution in a society of scarcity overlap with the problems and conflicts that arise from the production, definition, and distribution of techno-scientifically produced risks.[306]

The nature and impact of new risks in the contemporary differ significantly for him, including possible ecological hazards from new technologies like nuclear energy and genetic engineering. The contemporary late modern epoch that Beck characterizes as Risk Society is marked by difficulties in managing the new *side-effects* of modernization, as also the inability to govern the impact of new technologies through scientific control, be it through scientific knowledge or statistical probabilistic calculation of costs.

---

[304]   See Gabe Mythen, *Ulrich Beck* (Pluto 2004). See later in this section for his works on Risk Society.

[305]   Commentators position Beck's theorization as an intermediate position between the Marxist class analysis, Luhmann's functionalism and postmodern approaches. He is seen to combine his focus on new technological risk with the self-transformation of society through powerful processes of institutional individualization towards his thesis on social change in contemporary times, which he terms as late modernity. See for more: Jens O. Zinn, 'Risk Society and reflexive modernization' in Jens O. Zinn (ed.), *Social Theories of Risk and Uncertainty* (Blackwell 2008) 18, 19.

[306]   Ulrich Beck, *Risk Society: Towards a New Modernity* (Sage 1992) 19.

3.3.1 Reflexivity and Erosion of Scientific Monopoly

An important aspect for us is Beck's treatment of the breakdown of science's monopoly in defining risk and the resultant sharing of scientific and social rationalities during the ascription of safety in the contemporary epoch.[307] The contemporary is the third of the epochs in his categorization of history, viz., pre-modern, early modern and late modern.[308] In early modernity, scientific processes are understood and processed as an application of science to a given nature, people and society,[309] where scientific critique is organized and managed within science. Claims of truth production through science are largely unquestioned in spaces outside of organized science.[310] In his late modernity, there is scepticism among the general public towards accepting scientific and expert judgements without critical scrutiny. This is due to a number of reasons, including latent side-effects to health and the environment that are not predicted by science; adverse effects that are already felt by some social groups being not recognized or causally connected by the scientific establishment; the politics and economics of the production of scientific truth; and finally also due to the lessening of direct social or economic benefits to already affluent Western societies. The considerable time taken for recognizing the causal connections between various grave health concerns due to consumption of tobacco or use of various types of industrial asbestos are illustrative of the inadequacies of the scientific establishment and the accompanying public scrutiny of expert failures.

Beck used Brian Wynne's account of the continued use of the herbicide Agent Orange in the United States to elaborate this point. Scientific assertions of safety based on laboratory testing repeatedly failed in practice where the latent side-effects were systemically ignored by the scientific establishment. Scientific knowledge was seen to be produced in such a manner that it addressed the dangerous and harmful effects of the herbicide only through prescription of safe use based on laboratory conditions. The assumption was that the 'use of the herbicide offers no hazard to users, to the general publics, to domestic animals and to wildlife or to the environment generally, provided that the product is used

---

[307] Ibid 29.

[308] In the first period, natural disasters and hazards mark the horizon of societal understanding of risk, while the early modern period witnesses the advent of industrial hazards, where risks are perceived as calculable and manageable by financial compensation.

[309] Beck (n. 306) 154.

[310] Ibid 164.

as directed'.[311] Wynne demonstrated the problematic process through which this scientific assumption, based on laboratory conditions, was used time and again to reject various applications from the National Union of Agricultural and Allied Workers.[312] These applications sought a ban on the chemical based on substantial empirical evidence of grave health effects to workers accompanying the use of the chemical. Thus the confident scientific assertions of safety based on laboratory testing are persisted with, despite repeatedly failing in practice, since organized science can either demonstrate or claim a lack of hazard under laboratory conditions. This insistence has also purportedly fuelled the public scepticism about organized science. For the aforementioned reasons, Beck remarks that science has, either by its success or its widespread application, shifted to a reflexive phase, whereby it has to convince the public of the veracity of its claims. This contributes to a situation where it is no longer the scientific establishment that determines the validity of knowledge by itself, but it is dependent on societal acceptance and ethical compatibility.[313]

In Beck's treatment of latent side-effects, he is silent about an important political question in that they are neither latent nor peripheral for sections that are affected by them. The framing of what can be precisely termed latent or peripheral side-effects by the regulatory groups corresponds to the characterization of the future as risk as opposed to uncertainty or ambiguity.[314] For Beck, risks due to new technology become problematic because of factors of uncertainty, since there is never enough existing knowledge available from science and technology to control their occurrence or to deal with them through insurance.[315] This corresponds to the earlier mentioned characterization by Table of legislation of a range of incertitudes including uncertainty, ambiguity and ignorance, and the effect of the same on conventional science-based understanding of risk governance.[316] Irreversibility of the effects of new

[311] Brian Wynne, 'Frameworks of rationality in risk management towards the testing of naïve sociology' in Jennifer Brown (ed.), *Environmental Threats: Perception, Analysis and Management* (Belhaven Press 1989) 36.

[312] Ibid.

[313] Peter Weingart, *Anything Goes – rien ne va plus* (Kursbuch 1984) 61, 66, cited in Zinn (n. 305) 29.

[314] Chantal Mouffe, *On the Political* (Routledge 2005) 37–40.

[315] Ulrich Beck, *World Risk Society* (Polity 1999) 77. See also Ulrich Beck, *Ecological Politics in an Age of Risk* (Polity 1995).

[316] Text to nn. 277–279. See further, Andy Stirling, 'Risk, uncertainty and precaution: Some instrumental implications from the social sciences' in Frans

technologies, like, say, potentially to biodiversity in the case of GMOs, or catastrophic health effects on unborn babies in nuclear issues, make the effect of these incertitudes even more problematic. Further, there exist considerable amounts of complexity, contradiction, differentiation and fragmentation of scientific knowledge in various disciplines that have eroded scientific authority. He terms this contemporary condition 'reflexive modernity', which requires both science and public institutions to be reflexive. Thus this brings Beck to the truism that in Risk Society questions of true knowledge, especially in identifying risks, can no longer be answered only by science, but need further consideration. Though this does not mean that science becomes irrelevant, it leads to a competition between different scientific views, a buttressing of various political positions with often conflicting scientific expertise,[317] where faith, values and interests additionally need to be attended to, leading to a blurring of epistemic boundaries between science and politics.[318]

3.3.2 Organized Irresponsibility

The effects of eroding scientific rationality are particularly stark for the promise held by techno-scientific approaches to risk, because of the role of science in what Beck calls 'organized irresponsibility'. He elaborates on organized irresponsibility by seeking an account of 'how and why institutions of modern society unavoidably acknowledge the reality of catastrophe while simultaneously deny its existence, cover its origins and preclude compensation or control'.[319] The systemic avoidance of the latent side-effects by the scientific establishment in the earlier mentioned case of Agent Orange, provides a fitting foreground to his description of organized irresponsibility.[320] First, the institutional process of negotiating risks has become conflictual and contradictory, since institutional and legal forms that define risks at the same time legitimize and produce them.[321] The fragmentary nature of knowledge production, during the

---

Berkhout, Melissa Leach and Ian Scoones (eds), *Negotiating Change* (Edward Elgar 2003).

[317] Beck (n. 306) 167.

[318] Ibid 155–182; Jens O. Zinn, 'A comparison of sociological theorizing on risk and uncertainty' in Zinn (n. 274) 184.

[319] Ulrich Beck, 'Politics of Risk Society' in Jane Franklin (ed.), *The Politics of Risk Society* (Polity 1998) 9, 18.

[320] Text to n. 311.

[321] Beck, 'Risk Society' (n. 306) 29; Zinn, 'Risk Society and Reflexive Modernization' (n. 305) 27.

microcosmic focus of various disciplines, does not correspond to the complexities of risks that are unevenly distributed, and cumulate socially, spatially and temporally, whereby cause–effect relationships are particularly difficult to establish. Secondly, 'as long as a regulatory measure can be taken only when harm is proven beyond doubt, science supports and covers ongoing production of harms and dangers and the poisoning of the environment'.[322] Beck uses the example of toxic substances, where the list of banned substances always systematically lags behind the invention of new substances, as their use is allowed till proven to be unsafe, particularly since animal experimentation can be no substitute for effects on humans. Finally, the assumption that a specific level of final probability would imply safety legitimizes a high level of risk in practice. This is since the acknowledgement that acceptable levels of emission of harmful substances by single sources are misleading, as they accumulate over space and time. For instance, despite the existence of the stipulation in the recitals of the Deliberate Release Directive to study the risks of the cumulative and long-term effects of GMOs, the inability of risk assessors like EFSA to meaningfully incorporate them into their assessment has been strongly criticized.[323]

'Organized irresponsibility' is related to Beck's identification of a mismatch in Risk Society between the character of hazards and uncertainties related to hazards produced by late industrial society and its outdated 'relations of definitions'. Beck describes his 'relations of definitions' as the complex of rules, institutions and capacities that structure the conceptualization, identification and assessment of risk, and focuses on four such relations of definition in Risk Society. They are:[324]

1. Who is to determine the harmfulness of products or the danger of risks? Is it a responsibility that resides with those who generate those risks, with those who benefit from them or with the public agencies?
2. What kind of knowledge or non-knowledge about the causes, dimensions, actors, and so on, is involved? To whom does the *proof* have to be submitted?
3. What is to count as sufficient proof in a world in which we necessarily deal with contested knowledge and probabilities?

---

[322] Zinn (n. 305) 27.
[323] Text to nn. 197–198, n. 449.
[324] Beck, 'World Risk Society' (n. 315) 18–19.

4.   If there are dangers and damages, who is to decide on compensation for the afflicted, and on appropriate forms of future control and regulations?

These 'relations of definitions' are constituted by the legal, epistemological and cultural mix in which risk politics is conducted, where Beck asserts that we are trapped in a vocabulary that lends itself to an interrogation of the risks and hazards through the prior relations of definition of simple, classic first modernity. He seeks to understand the shifts in these four relations of definitions, where the increase in obviousness of the nature and degree of danger is accompanied by a slip through the nets of proofs, attributions and compensation with which the legal and political systems have attempted to deal with these dangers.[325] The nature of these relations of definitions is fundamental since they have profound impacts on how we organize our societies. Decisions like appropriate forms of compensation, or the question of who has the responsibility to ascribe harmfulness of a substance (the producers of the substance that profit from it or the public agencies) are intensely political, necessitating political resolutions.

Beck notes that the knowledge for assessing contemporary risk is not restricted to scientific establishments, but also is situated among lay people, citizen groups, civil society organizations and other social institutions, often mediated by the media. Though scientific foundation is often used by a range of social actors, social impact of risk definitions cannot be determined by scientific evidence and provability, but only through public engagement.[326] This is also since people are acutely aware of expert disagreements and economic interests of the scientific-industry establishment, which is also recognized to produce risk in the process of generation of technology.[327] Beck identifies the particular breakdown of this rationality in risky technologies like nuclear and biotechnologies, since these technologies have to be built or developed for their functioning and risks to be studied, and where mistakes in this activity can mean that reactors can leak or explode, people get killed, contract BSE ('mad cow disease') and so on, where society becomes a laboratory, where there is no one responsible for its outcomes.[328]

---

[325]   Beck, 'Ecological politics' (n. 315) 116–118, 129–133, 136–137.
[326]   Beck, 'Risk Society' (n. 306) 32.
[327]   Tulloch (n. 274) 146.
[328]   Beck, 'Politics of Risk Society' (n. 319).

Therefore, not only is the monopoly of rationality of science already undermined in this account,[329] but it also reflects the dangers often posed by techno-science itself. Beck underlines the ambivalent position of science in society, instead of the earlier image of science as the motor for economic growth and healthy well-being. Thus, in situations of risk conflicts, politicians can no longer reasonably rely on scientific experts, since there are always competing and conflicting claims about the definition of risk in every concrete context of risk conflict, where conflicting knowledge cannot be a matter of good and bad experts. Experts can supply more or less uncertain factual information about probabilities, but can never answer which risk is acceptable and which is not. At the same time, various groups use science to make their arguments more persuasive and to immunize ideologies and interests with scientific rationalities.

Further, experts and laypeople may refer to different risks: while experts make a balance of the gains against possible deaths in the normal production process, lay people regularly refer to the impacts of possible accidents even though they are highly unlikely.[330] The nature of applicability of scientific advice becomes a question that wider society has to answer in broadly ethical and political terms, since a mere implementation of scientific advice locks society into the mistakes, modes and uncertainties. Beck, thus, reasons that politics and morality have to gain priority over shifting scientific reasoning. Following this, new reasonable strategies to manage uncertainties and ignorance are required in Risk Society. This is because in the modern strategy to control the future rationally through risk calculation, the acceptance and reliability of the forms of statistical-probabilistic calculation are eroded, and faith and values are additionally needed to gain social trust. Given the fact that we are dealing with the problem of risk produced through specific technological trajectories, say as opposed to natural disasters,[331] the central dilemma is about deciding which kinds of risks society is willing and able to take. Additional intellectual resources are necessary to understand how risk has become a matrix in these political decisions, be it in the important relations of definitions or which kinds of risks society is

---

[329]   Beck, 'Risk Society' (n. 306) 57.

[330]   Starr (n. 267).

[331]   Particularly since one has to acknowledge that even the securest technology sometimes fails, and accidents are *normal* even in allegedly safe technologies: Charles Perrow, *Normal Accidents: Living with High Risk Technologies* (Princeton University Press 1984).

willing to take, which are areas in which science can no longer have a monopoly in generating truth.

3.4   RISK IN GOVERNMENTALITY SCHOLARSHIP

Governmentality scholarship conceptualizes risk as being part of a specific technique to govern societies, which developed in liberal states in western industrialized Europe during early and later modernity.[332] As an analytical technique, Governmentality seeks fault-lines in the suppositions of stability in the contemporary, towards revealing its contingency. Its foundations are generally attributed to Foucault's 1978 essay on Governmentality that discusses the change in rationality of modern governments from seventeenth century Europe.[333]

> The focus was to be on destabilizing and questioning the present by revealing its contingent formation, its non-necessity. In governmentality, emphasis would be on how that which appears as necessary is to be understood as assembled together out of available materials, ideas, practices, and so on, in response to a specific understanding of the nature of the problems to be solved. In tandem with this, emphasis was placed on the understandings and constructions of the world that give rise to efforts to change it. In this view, that which appears natural is not to be taken for granted as something – like 'population' or 'the economy' – unproblematically real and just waiting to be discovered. Rather, it is to be regarded as invented, reflecting or embodying governmental *understandings* of the way things are. In the same moment, as the arbitrariness of many taken-for-granted categories in the present is made visible, possibilities for change emerge – the analytic gives rise to insights into how things might have been otherwise, and thus how they *could* be different in the future.[334]

The argument is that Governmentality or governmental rationalities are always linked to the techniques of answering the question 'what is to be done?'[335] 'Welfare liberalism, the governmental rationality that dominated much of the twentieth century in Europe, was linked to all manner

---

[332]   Michel Foucault, 'Governmentality' in Graham Burchell, Colin Gordon and Peter M. Miller (eds), *The Foucault Effect: Studies in Governmentality, with two lectures by and an interview with Michel Foucault* (Harvester Wheatsheaf 1991) 87.

[333]   Ibid.

[334]   Pat O'Malley, 'Governmentality and risk' in Jens O. Zinn (ed.), *Social Theories of Risk and Uncertainty* (Blackwell 2008) 52, 54.

[335]   See Jacques Donzelot, 'The power of political culture' (1979) 5 *Ideology and Consciousness* 71.

of practical technologies, such as risk-based schemes'.[336] Various scholars including Defert, Ewald and Castel used the Governmentality approach to examine the contemporary governance tool of risk; albeit prominently in specific realms of risk technologies of insurance and psychiatry.[337]

Dean attacks the over-generalization in Risk Society theorization, and its emphasis on incalculability of risk, when one takes into account that risk is always calculated and represented in an attempt to construct coherent programmes of government in the face of uncertainty, even though such calculation may be ad hoc, likely to fail and primarily taken in order to be seen to be doing something.[338] Risk within Governmentality scholarship is not regarded as intrinsically real, but is seen as a very abstract probabilistic tool, a particular way through which a problem is imagined and dealt with. While a large number of events are sorted into a distribution towards making probabilistic predictions, particular details of each case are submerged or stripped away for a complex assemblage of elements to achieve some ends. Within this scholarship, the point is not to evaluate whether such procedures are fair or accurate, but rather to analyse the specific characteristics that underlie attempts at governing uncertain future events.[339] Risk is therefore understood not as harm or danger, but as a specific way to manage such threat with the help of calculative technologies. 'A harm or danger just happens; it only becomes a risk when it is brought into being as a probability of an event, also usually attributed to a specific reason'.[340] Further, risk is seen as part of a societal discourse that produces the knowledge or truth assertion to define reasonable action and decision-making, linking it to normative and

---

[336] O'Malley (n. 334) 56.

[337] See Daniel Defert, '"Popular life" and insurance technology' in Graham Burchell, Colin Gordon and Peter M. Miller (eds), *The Foucault Effect: Studies in Governmentality, with two lectures by and an interview with Michel Foucault* (Harvester Wheatsheaf 1991) 211; Francois Ewald, 'Insurance and risk' in Graham Burchell, Colin Gordon and Peter M. Miller (eds), *The Foucault Effect; Studies in Governmentality, with two lectures by and an interview with Michel Foucault* (Harvester Wheatsheaf 1991) 197; Robert Castel, 'From dangerousness to risk' in Graham Burchell, Colin Gordon and Peter M Miller (eds), *The Foucault Effect; Studies in Governmentality, with two lectures by and an interview with Michel Foucault* (Harvester Wheatsheaf 1991) 281.

[338] Mitchell Dean and Barry Hindess, *Governing Australia: Studies in Contemporary Rationalities of Government* (Cambridge University Press 1998) 145; Tulloch (n. 274) 153.

[339] O'Malley (n. 334) 54.

[340] Zinn (n. 318) 174–175.

moral issues in society in a fundamental way.[341] Where, even in cases where risk is introduced as evidence-based, whether and how such objectivist calculations are applied and which criteria are selected are necessarily confounded with moral judgements. O'Malley stressed that differing approaches to risk have in turn given shape to different forms of liberal government.[342] Here, different applications of calculative technologies like insurance risk, epidemiological risk, case management risk and clinical risk are focused upon to understand how these technologies are embedded and connected to social sense-making. Most of the Governmentality scholarship has focused more on actuarial and psychiatric risk, rather than, say, epidemiological risk, which is far more applicable in the regulatory experience in risk assessments of GMOs. However, 'epidemiological risk is similar to insurance risk in that the calculus of risk is undertaken on the basis of a range of abstract factors and their correlation within populations ... though, it has its own discursive rationality and set of techniques and interventions'.[343] Unlike in the realm of insurance, it is not the loss of capital but health outcomes of populations that are subject to risk calculation in epidemiological risk. The protection of investment capital in terms of goals like protection of innovation, techno-scientific development for economic growth and well-being, defined as public interest, are values that could come into the picture in the frames of risk calculation here.[344]

Though the patterns of technological risk are less explored in Governmentality scholarship two relevant factors stand out. First, Dean seeks to argue that in epidemiological risk the attempt is to reduce projections of risk, and individuals are not centred at all in these projections, using statistical and probability modelling.[345] Secondly, like Risk Society scholarship, Governmentality researchers have also drawn attention to the importance placed upon the self-management of risk and the increasing privatization of risk.[346] Both these aspects appear to be relevant in an examination of the EU risk regulation of GMOs. For instance, in the

---

[341] Tom Baker and Jonathan Simon (eds), *Embracing Risk: The Changing Culture of Insurance and Responsibility* (University of Chicago Press 2002); Richard Victor Ericson and Aaron Doyle (eds), *Risk and Morality* (University of Toronto Press 2003).

[342] Pat O'Malley, *Risk, Uncertainty and Government* (Glasshouse 2004).

[343] Mitchell Dean, *Governmentality: Power and Rule in Modern Society* (Sage 1999) 143.

[344] Castel (n. 337).

[345] Dean, 'Power and Rule' (n. 343).

[346] Beck, 'World Risk Society' (n. 315) 4.

previous chapter we saw that data about potential hazards are generated, collated, located and analysed by the private applicants interested in their market use, right from the start of the development of this technology. The starting point of the regulation of GMOs under the Contained Use Directive is a dossier that is prepared by the applicant who is pursuing the research and development of specific GMOs within laboratories and greenhouses. The basis of classification for the purpose of containment and danger is based on this assessment.[347] The importance of the dossier, which provides the field data in the decision for field trials under the Deliberate Release Directive forwarded by private parties developing the GMOs, is pivotal to the subsequent statistical analysis by EFSA and Member State authorities.[348] Similarly, the base date used for statistical risk analysis before authorization of release under the Deliberate Release Directive and the Food and Feed Regulation is also generated by the developers, who are interested in the swift authorization of their application to release or cultivate.[349] In each of these instances, the private party is responsible for observing and collecting relevant data, whereby the generation and control of the initial steps are always with them. They are also responsible for analysing the data and sharing their risk analysis with the public authorities like EFSA, and to a limited extent with the general public.[350] Thus, in this movement of key information through different phases of risk regulation that ends in formal risk analysis, it is only at the very last stage of formal risk analysis that a formal transfer of data from the individual operator to a public body like EFSA is required. Such data, which are so pivotal in the risk analysis and its techniques of statistical predictions, are fully generated and collated under the control of the developer. This aspect is further complicated in jurisdictions like the US, where the system of end user agreements has been used to stifle public scientific enquiries. Here, technology stewardship agreements generally accompany most seed sales that, inter alia, have forbidden buyers from conducting even non-profit research on the seed without prior permission from the patent-owning corporation. Independent scientists who might feel that appropriate environmental and food safety tests have not been conducted before regulatory approval for the seed will not

---

[347] Text to nn. 171–173.
[348] Text to nn. 177–178.
[349] Text to nn. 148–149.
[350] Text to n. 155.

be able to conduct them on their own, without prior industry approval.[351] In this context, critics in the US cite their personal experience, to assert that 'only studies that the seed companies have approved ever see the light of a peer-reviewed journal'.[352] In a number of cases, 'experiments that had the implicit go-ahead from the seed company were later blocked from publication because the results were not flattering'.[353] What is the nature of proof that is employed here, to borrow Beck's relations of definitions, which allows the same person to generate the knowledge (which is the basis of subsequent statistical projection by public agencies), as well as own and profit from it? The challenge for law is to ensure a safety regime that is not affected by such convergence of institutional locations that generate data that are used to model prediction of danger as well as generate profit. It would appear that such a convergence can only be legitimated by active building of trust in these mechanisms, rather than by an assumption of passive trust in expert institutions, based on traditional notions of legitimacy and classical representation that assume experts as representatives of the general public, and expert rationality as spokesperson for public judgement.

A second line of enquiry regards the nature of techniques through which the side-effects of the use of GMOs temporally, spatially and socially accumulate. Despite the stated regulatory objective of a reduction in epidemiological risk to the environment and human health, these temporal, spatial and social accumulations are ignored in the statistical projections of risk analysis. While there is a professed emphasis on reduction of risk as rectification for a general cultural expectation of safety,[354] the process through which a specific probabilistic model and level of risk can legally imply safety needs further understanding. A specific level of risk that is found to be acceptable in a regulatory system is related to the rationality and normativity employed in it, a level that some sections of the public may find too low. The type of public

---

[351] Emily Waltz, 'Under wraps'(2009) 10 *Nature Biotechnology*, 880, <http://www.nature.com/nbt/journal/v27/n10/full/nbt1009-880.html> last accessed 10 November 2011.

[352] Editorial, 'Do seed companies control GM crop research?' (2009) *Scientific American Magazine* (August, published 21 July) <http://www.scientific american.com/article.cfm?id=do-seed-companies-control-gm-crop-research> last accessed 10 November 2011.

[353] Ibid.

[354] Frank Furedi, *Culture of Fear: Risk Taking and the Morality of Low Expectation* (Cassell 1997) 3, 'the contemporary is dominated by a peculiarly defensive attitude to risk, in which risks are almost always framed by the precautionary principle of better safe than sorry'.

participation needed to make these choices in an acceptable and trust-worthy manner needs wider deliberation through law. In recent times, the precautionary principle has been professed as a logic and norm to regulate safety concerns of technologies in general to enhance public trust and reliability. Before we attempt a detailed examination of the principle it is important to underline the importance of the stated values of trust and engagement in the employment of the tool of risk and articulations of safety.

3.5 TOWARDS TRUST AND ENGAGEMENT

If risk is not an artefact that is discovered or measured through expert meditation, but is understood as an ascriptive tool that makes the issue of safety governable, then trust and engagement are to be seen as values that are desired by law. This is keeping in mind the unknowns and uncertainties intrinsic to the risk assessment of emerging technologies,[355] 'as opposed to scientists as experts on scientific truths, as truth speaking to power in a traditional picture, in risk assessment scientists enter the public arena as experts who are part of a complex rhetoric and political system'.[356] Trust and engagement are increasingly seen as tropes through which anxieties accompanying techno-scientific progress are mediated to further dominant rationalities. In the event of intense risk conflicts, such a political system would require an emphasis on these values through various levels and types of public engagement.[357] These values have been recognized as key features of any risk regulation system 'if it is to work'.[358] At the same time, commentators notice the wide discrepancy regarding the professed acceptance of social, political and ethical considerations by mainstream regulators that 'rarely feed through into decision-making'.[359] This is since the 'broadly conceptualized concerns

---

[355] John Kadvany, 'Taming chance: Risk and the quantification of uncertainty' (1996) 29 *Policy Sciences* 1, 10.

[356] Karen Kastenhofer, 'Risk assessment of emerging technologies and post-normal science' (2011) 36 *Science, Technology and Human Values* 307, 326.

[357] Christopher Hood, 'Where extremes meet: "Sprat" versus "shark" in public risk management' in Christopher Hood and David Jones (eds), *Accident and Design: Contemporary Debates in Risk Management* (UCL Press 1996).

[358] Brian Wynne, 'Risk and social learning: Reification to learning' in Sheldon Krimsky and Dominic Golding (eds), *Social Theories of Risk* (Westport 1992).

[359] Maria Lee, 'Beyond safety? The broadening scope of risk regulation' (2009) 62 *Current Legal Problems* 242.

about the social or ethical commitment embedded in a technology' are sidestepped by turning to 'technical expertise on "safety" and "risk"'.[360]

Given organized techno-science's fractured relationship of trust with wide sections of the public and the related credibility issues, as well as the power and information imbalances between techno-scientific realms and other sections of society, the problems with classical representation models in technology regulation are clear. In the EU GMO regime, meaningful public participation that actively builds public trust and political engagement is seen as a pertinent way to approach this problem. This recalls Luhmann's category of risk communication, and his second order differentiation between risk and danger, where those who feel they are in control of a decision consider a dangerous event arising from a decision as risk, while others may consider the risk regulation itself as a source of danger. In principle, the core question concerns the kind of process law has to imagine for mediating and engaging with the question of attribution. If this should be a political process, the contours of possible engagement between the decision-maker and the outsider, including the opposition to a technology, become central. What tools of engagement could make the outsider feel part of a decision, and build the kind of trust to bridge this gap? The concern for exploring conceptual and institutional arenas necessary for such a shift leads us to enquire about the room provided by EU legislation for public participation in GMO regulation.

As argued in this chapter, issues of trust do not necessarily pertain to the normativity of science, but to the statistical projections and models through which the occurrence of hazards is reduced to probabilities in techno-scientific characterizations of risk. These evaluations will surely have to include sufficient competence in technical investigations about dangers due to the use of a technology. Towards this end, two kinds of trust need attention: first scientific/technical processes should be honed towards identifying and investigating all types of potential hazards. Such potential hazards may include some that are not considered particularly relevant by techno-scientific communities, but are required due to the need for wider engagements towards an active pursuit of social trust. Secondly, trust for governance and regulation around risk needs wider engagement than mere investigations by techno-scientific communities. Though the modalities and norms of this engagement require more elaboration, it would include a more public estimation of the social value and a comparative appraisal of alternate trajectories that have similar

---

[360] Ibid.

social functions. In other words, the challenge relates to the movement from an imagination of passive trust in classical risk mechanisms (where trust is invested mainly in expert bodies as part of the conceptualization of delegation within formalistic liberal democracies) towards generating institutional rationalities for active trust and democratic validation through programmes of public engagement and participation. The nature and contours of such programmes need broader deliberation, including the challenge of nurturing high levels of competence and scientific research about dangers in new technologies during these endeavours.

To sum up, by engaging with the sociological descriptions of risk and identifying the serious limitations of techno-scientific conceptualization of risk, this chapter has aimed at understanding law's employment of risk as a tool of governing technology. Given the assumptions of post-normality in science, and where risk is understood as ascriptions to make questions of safety governable, public participation is argued as central for an appropriate characterization and employment of risk. To reiterate the point, usually public participation mechanisms are seen as instruments to make pre-existing construction of risk more acceptable to various sections of society. However, this chapter has sought to demonstrate that contemporary sociological understandings put public participation at the heart of the construction of risk, and not as an adjunct to make expert constructions more acceptable. If this is so, newer institutional and conceptual arenas that facilitate engagement between expert systems and other sections need to be imagined. Explorations of appropriate frames to answer these questions require a further examination of the safety regime for releases of GMOs in Europe. Continuing from the previous chapter, where the framework of the EU GMO regulation was identified, the following chapter examines the nature and contours of employment of the precautionary principle. The principle is often asserted as the cornerstone of the GMO regulatory framework, and has had a peculiar relationship with the concept of risk, central to cultural demands of safety, and the connected issue of trust. Therefore its employment in EU GMO regulation requires a detailed examination.

4. Precaution, public participation and technocratic responses

> How precaution can function in a debate is doubly speculative: the future science and technology is still uncertain, let alone the future world in which these may function and have effects.[361]

The precautionary principle has been asserted as a cornerstone in the risk regulatory framework for GMOs in Europe. As suggested in this chapter, invocations of the principle in many regulatory realms often appear to be geared towards fostering public confidence and trust, either explicitly or covertly, and for regaining legitimacy for a food safety system battered by controversies like BSE. In view of the identification in the previous chapter of trust and engagement as key features of such systems that ascribe safety, we now take a closer examination of EU GMO regulation to understand the implications of a declared emphasis on precaution. Given constructions of classical risk and precaution as emerging from different paradigms, this chapter seeks to understand if and how improvements have been made to take public disaffections about statistical models, used for ascribing probabilities, which underpin a classical mode of risk governance, seriously. This investigation about any substantial improvement based on the precautionary principle includes how the current regime is more responsive to earlier identified norms of trust and engagement in ascriptions of safety, including in attempts at public engagement and participation.

Originally intended to provide guidance in environmental cases where societal stakes are high and scientific claims uncertain or complex, the principle had emerged as 'the hippest environmental principle on the global block' by the turn of the twenty-first century.[362] Though conventional histories trace the principle from German administrative law

---

[361]   Rip (n. 23) 273.
[362]   Elizabeth Fisher, 'Evolution and status of the precautionary principle in international law' (2002) 4 *Environmental Law Review* 258.

principles in *Vorsorgeprinzip*,[363] simultaneous and equally lengthy histories of precautionary-type policies in many other jurisdictions are amply
documented.[364] Termed as a sobering rejoinder to the overzealous promotion of 'sound science' in public policy,[365] the principle is now part of
law and policy in a number of jurisdictions across most parts of the
world.[366] An inordinate breadth of literature regarding (a) the range of
divergence in the employment of the principle in national, international
and regional legal instruments,[367] (b) its substance and meaning,[368]

---

[363]   Mags D. Adams, 'The precautionary principle and the rhetoric behind
it' (2002) 5 *Journal of Risk Research* 301. She underlines the history of the
principle in German environmental policy, and identifies it as one of the five
fundamental principles underlying German environmental policy. She transliter-
ates the precautionary principle as a 'before care' principle, or the principle of
taking care before we act. See further: Konrad von Moltke, 'Three reports on
German environmental policy' (1991) 33 *Environment* 25.

[364]   See for instance Poul Harremoës (ed.), *The Precautionary Principle in
the 20th Century: Late Lessons from Early Warnings* (Earthscan 2002); Thomas
McGarity and Sidney Shapiro, *Risk Regulation at Risk: Restoring a Pragmatic
Approach* (Stanford University Press 2003); Hans Jonas, *The Imperative of
Responsibility* (University of Chicago Press 1984).

[365]   Theodore Porter, *Trust in Numbers: The Pursuit of Objectivity in
Science and Public Life* (Princeton University Press 1995).

[366]   By the mid-1990s the principle was already seen as increasingly
integrated into international treaties as a foundation for environmental decision-
making: David Freestone and Ellen Hey, *The Precautionary Principle and
International Law: The Challenge of Implementation* (Kluwer 1996). See for the
significant presence of the principle in the European Union: Nicolas de Sadeleer,
'The Precautionary principle in EC health and environmental law' (2006) 12
European Law Journal 139; in the erstwhile British commonwealth countries:
Jacqueline Peel, 'Interpretation and application of the precautionary principle:
Australia's contribution' (2009) 18 *Review of European Community and International Environmental Law* 11; in many African countries and African regional
legal instruments: John S. Applegate, 'The taming of the precautionary principle'
(2002) 27 *William and Mary Environmental Law Review* 13. Though conventional understanding is that precaution is not favoured in the US due to its
opposition to the principle's application in international law, and also not having
officially adopted it, there are ample areas of precautionary regulation in the US,
where it could be considered even more precautionary than in the EU. See for
instance: Jonathan B. Wiener and Michael D. Rogers, 'Comparing precaution in
the US and Europe' (2002) 5 *Journal of Risk Research* 317; Stirling and van
Zwanenberg (n. 289).

[367]   See for a useful survey of literature: Arie Trouwborst, 'The precautionary principle in general international law: Combating the Babylonian confusion'
(2007) 16 *Review of European Community and International Environmental Law*
185, fn 3.

(c) normative and practical implications of the principle,[369] (d) plurality of related terminologies[370] elicited a commentator's quip that academic attention has spent rivers of ink on precaution.[371]

Parallel to this divergence in the substance, meaning and implications of the principle, precaution has also become a ubiquitous feature in regulatory architectures pertaining to environmental and health issues, and even beyond, in other important regulatory realms.[372] The principle is increasingly exhorted as a guiding principle of EU law in various fundamental ways, and has been an established feature in EU risk regulation regimes over the past two decades. Article 114 of the TFEU stipulates Community policy to be based on the precautionary principle while aiming at a high level of protection for the environment. The integration principle (enunciated in Article 11 TFEU) implies that the precautionary principle ought to apply to the development and implementation of all environmental initiatives regardless of the particular policy sphere in which they occur.[373] The principle has been particularly important in health, environment and agricultural sectors of the EU regulatory policy,[374] and in the way risk and other standards of social

---

[368]   Ronnie Harding and Elizabeth Fisher (eds), *Perspectives on the Precautionary Principle* (Federation Press 1999); Timothy O'Riordan and Andrew Jordan, *The Precautionary Principle, Science, Politics and Ethics* (Centre for Social and Economic Research on the Global Environment 1995).

[369]   Cass Sunstein, *Laws of Fear: Beyond the Precautionary Principle* (Cambridge University Press 2005); Luciano Butti, *Implementing the Precautionary Principle: Approaches from Nordic Countries and the EU* (Earthscan 2007); Timothy O'Riordan and James Cameron (eds), *Interpreting the Precautionary Principle* (Earthscan 1994); Timothy O'Riordan, J. Cameron and Andrew Jordan (eds), *Reinterpreting the Precautionary Principle* (Cameron May 2001).

[370]   See further, Jacqueline Peel, 'Precaution – A matter of principle, approach or process?' (2004) 5 *Melbourne Journal of International Law* 1.

[371]   Jaye Ellis, 'Overexploitation of a valuable resource? New literature on the precautionary principle' (2006) 17 *European Journal of International Law* 445.

[372]   See for a general overview: The World Commission on the Ethics of Scientific Knowledge and Technology, *Expert Group Report on the Precautionary Principle* (COMEST 2005) <http://unesdoc.unesco.org/images/0013/001395/139578e.pdf> last accessed 20 December 2011.

[373]   See for an appropriate elaboration: Joanne Scott, 'The precautionary principle before the European Courts' in Richard Macrory, Ian Havercroft and Ray Purdy (eds), *Principles of European Environmental Law* (Europa Law Publishing 2004) 51, 54.

[374]   Regulation 1907/2006 of the European Parliament and of the Council concerning the registration, evaluation, authorization and restriction of chemicals

appropriateness regarding safety of new technologies are adjudicated by EU courts.[375] The Commission identified the principle as a 'key tenet of its policy' not only in the realm of the environment but also for human, animal and plant health.[376]

The quest for public trust in risk regulation through the invocation of the principle is discernible in Member States' policy documents. For instance, the UK Department for Environment, Food and Rural Affairs (DEFRA) aspires to the 'gain [of] public trust in the policy areas for which we have responsibility by following and communicating a precautionary approach – we will not always wait until we have proof of a potential threat and its impact before taking action or issuing advice'.[377] The Dutch Scientific Council for Government Policy termed precaution as a constitutional task in an attempt to 'future proof' the safety policy framework.[378] The Council suggested that the policy framework has to be developed within a new paradigm that starts from the normative perspective of the precautionary principle, requiring a proactive approach to uncertainty and vulnerability of people, society and the natural environment.

It would appear that the explicit invocation of the precautionary principle in the current GMO regime (dealt with in a later section) was absolutely necessary for those sections that were sceptical of the existing 'science-based' release mechanism. During its introduction, the current regime was hailed as the most rigorous assessment framework in the world, because of its precautionary assertions: 'one that will reinforce international credibility, increase public confidence, and address the most critical concerns of the environment and human health'.[379]

[2006] OJL31/1; Directive 2000/60/EC of the European Parliament and of the Council of establishing a framework for the Community action in the field of water policy [2000] OJ L327/1; Directive 92/43/ EEC of the Council on the conservation of natural habitats of wild fauna and flora [1992] OJ L 206/7.

[375] Text to nn. 468–478.

[376] European Commission, 'Communication from the Commission on the Precautionary Principle' COM (2000) 1 final.

[377] DEFRA, *Risk Management Strategy* (April 2002) para. 2.9.

[378] WRR (n. 2) 161–162.

[379] 'We address the most critical concerns of public [sic] regarding the environmental and health effects of GMOs and enable the consumers to choose ... European consumers can now have confidence in that any GM food or feed marketed in Europe has been subject to the most rigorous pre-marketing assessment in the world'; Commission Press release, 'European Legislative framework for GMOs in place' (IP/03/1056, 22 July 2003).

4.1   CORE CLAIMS AND THEIR CRITIQUES

While avoiding an exercise of 'precaution-spotting',[380] it is important to recognize at the outset that substantive versions of this principle vary greatly in various international[381] and EU legal instruments.[382] 'Deeply ambivalent and apparently infinitely malleable', the principle, depending on one's perspective, is either asserted as 'a significant step towards better environmental regulation' or derided as 'a dangerous and meaningless principle that threatens to destroy industrial economies'.[383] Notwithstanding different formulations, four elements stand out among commentaries as most relevant to make sense of the principle:[384] the trigger for the invocation and use of the principle,[385] the timing,[386] the nature of the regulatory response,[387] and iteration.[388]

---

[380]   See for the use of such a term to connote a widespread research practice of writing about the principle through its identification in a broad range of legal instruments: Elizabeth Fisher, 'Precaution, precaution everywhere: Developing a "Common Understanding" of the precautionary principle in the European Community' (2002) 9 *Maastricht Journal of European and Comparative Law* 7.

[381]   See Arie Trouwborst, *Evolution and Status of the Precautionary Principle in International Law* (Martinus Nijhoff 2002).

[382]   See for a useful elaboration: Eloise Scotford, 'Mapping the Art. 174 (2) EC case law: A first step to analyzing community environmental law principles' (2008) 8 *Yearbook of European Environmental Law* 1.

[383]   Joanne Scott and Ellen Vos, 'The juridification of uncertainty: Observations on the ambivalence of the precautionary principle within the EU and the WTO' in Christian Joerges and Renaud Dehoussse (eds), *Good Governance in Europe's Integrated Market* (Oxford University Press 2002) 253.

[384]   Cited in Applegate (n. 366). See also Per Sandin, 'Dimensions of the precautionary principle' (1999) 5 *Human and Ecological Risk Assessment* 889, 890–895.

[385]   The characterization or nature of understanding of the suspected hazard with respect to the degree and possibilities of harm. An important element here is of course the appropriateness of the nature of understanding and source of the information that is used to characterize this suspicion.

[386]   The nature of the relationship between the trigger and the taking of regulatory action in terms of time: at what stage are appropriate actions permitted or even required, and in some cases at what stage are uncertainties deemed to be resolved.

[387]   Regulatory actions can range from a requirement for an outright ban to a very high flexibility in the manner in which risk is sought to be avoided, including an encouragement for a far more focused/strong/serious research, monitoring, and so on.

[388]   Since action taken is based on the premise of uncertainty, does the particular version require further action when reasonable amounts of certainty are

Principle 15 of the Rio Declaration is often quoted as an authoritative formulation of the principle:

> in order to protect the environment, the precautionary approach shall be widely applied by States according to their capabilities. Where there are threats of serious or irreversible damage, lack of full scientific certainty shall not be used as a reason for postponing cost-effective measures to prevent environmental degradation.[389]

The core of the principle lies in the recognition of the need for avoidance and minimization of harm to human health and the environment through anticipatory and preventive regulatory controls; it is often conceptually asserted that the principle embodies notions of caution, care, preventive action, common sense and prudential responsibility.[390] To this end, there is an expectation to put regulatory restrictions (that are not necessarily a ban) in place, on activities and technologies whose environmental and health consequences are uncertain and potentially serious, at least until the uncertainty is more or less resolved.

Reviewing the evidence of early warnings of the devastating effects of regulatory responses to the use of particular technological products that were initially publicized as wonderfully advantageous for society, and the social and economic cost of these regulatory lapses, David Gee and Morris Greenberg noted:

> [L]ong term environmental and health monitoring rarely meets the short-term needs of anyone, thus requiring particular institutional arrangements if it is to meet society's long term needs...the early warnings of 1898–1906 in the UK and France were not followed up by the kind of long-term medical and dust exposure surveys of workers that would have been possible at the time, and which would have helped strengthen the case for tightening controls on dust levels. Even now, leading asbestos epidemiologists can conclude it is unfortunate that the evolution of the epidemic of asbestos induced mesothelioma, which far exceeds the combined effects of other known occupational industrial carcinogens, cannot be adequately monitored.[391]

---

regained, and, if so, what is this level of 'reasonable', and what further action is required to facilitate the reaching of this stage.

[389]   <http://www.unep.org/Documents.Multilingual/Default.asp?documentid= 78&articleid=1163> last accessed 10 November 2011.

[390]   Jane Hunt, 'The social construction of precaution' in O'Riordan and Cameron (n. 369) 172.

[391]   David Gee and Morris Greenburg, 'Asbestos: From magic to malevolent minerals' in Harremoës (n. 364) 52.

This characterization points to scenarios where expert communities invariably fail to pick up early warnings (regarding new technological products that are hailed by their proponents as wonderful solutions to fundamental problems) emanating from other sections of the public who experience and engage with technology differently. This underlines the importance of development of new institutional arrangements to overcome this difficulty, for the effective realization of safety concerns.

In a significant academic collaboration that provides an overview of the implementability and applicability of the principle,[392] four major themes were identified as widely considered essential for an appropriate understanding of the principle. First is a thematic focus on the widespread implications of implementation of the principle for decision-making institutions and processes. Here the principle is intended to serve as a basis for deliberation at different levels of decision-making, in contrast with classical risk regulation, and in particular the appropriate role of science in decision-making.[393] Secondly, a necessary emphasis on the legal and regulatory cultures in which the principle operates makes the description and understanding of the principle even more complicated. It is therefore necessary to move beyond conventional descriptions of the principle, say in terms of a shifting of the burden of proof or as an adjustment to the standard of proof, since existing institutional landscapes are far more varied, making such rudimentary remarks of limited use. Thirdly, particular environmental and public health problems create specific challenges, in particular regarding the question of scientific ignorance, viz., 'we don't know what we don't know'.[394] The fourth theme is of deliberation, and explores the close connection between application of the principle and deliberative practices outside the confines of formal institutional democracy, as well as civil and political society. Once deliberation is understood in a broader sense – that is, as free and public reasoning among equals concerned with the weighing of pros and cons in relation to particular choices, options and measures of public concern with a view to decision-making – the institutional forms in which deliberation can take place and the type of communication that is necessary in particular participatory processes become inextricably connected with the employment of the principle.

---

[392] Elizabeth Fisher, Judith Jones and Rene von Schomberg (eds), *Implementing the Precautionary Principle: Perspectives and Prospects* (Edward Elgar 2006).

[393] Elizabeth Fisher, Judith Jones and Rene von Schomberg, 'Implementing the precautionary principle: Perspectives and prospects' in ibid 1, 8–9.

[394] Ibid 9.

Both substantive and procedural rationalities coexist in the quest for better governance of safety concerns through precaution:

> Here substantive rationality is in terms of providing a more appropriate decision-making tool to deal with situations of scientific uncertainty, ignorance and lack of identified (yet proven) causality of identified potential hazard/harm. Procedurally, the principle facilitates the communication between risk assessors, risk managers and the public in a potentially enabling way of democratic societal choice about the level of acceptable risk, as well as the procedure for ascertaining it.[395]

Varying possibilities of identification and characterization of threats, and degrees of certainty, provide windows for a range of rationalities and worldviews in the implementation of the principle. Justification for a particular kind of substantive rationality of the principle in the EU legal framework is well captured in von Schomberg's formulation:

> Where, following an assessment of available scientific information, there are reasonable grounds for concern for the possibility of adverse effects but scientific uncertainty persists, provisional risk management measures based on a broad cost–benefit analysis whereby priority will be given to human health and the environment, necessary to ensure the chosen high level of protection in the Community and proportionate to this level of protection, may be adopted, pending further scientific information for a more comprehensive risk assessment, without having to wait until the reality and seriousness of those adverse effects become fully apparent.[396]

Here, at the very least, the principle empowers public decisions attempting to prevent harm in situations of possibilities of hazards against the argument of lack of full scientific certainty. It includes justification for wider participation and active public engagement in investigations about concerns of safety and appropriateness of technological trajectories, societal needs and pathways of development. The concept of scientific uncertainty here bridges the substantive and procedural rationalities of the principle, where often an acknowledgement of scientific uncertainty

---

[395]   Laurence Boisson de Chazournes, 'New technologies, the precautionary principle and public participation' in Theresa Murphy (ed.), *New Technologies and Human Rights* (Oxford University Press 2009) 161, 177–179. See also Joel A. Tickner and Sara Wright, 'The precautionary principle and democratizing expertise' (2003) 30 *Science and Public Policy* 213.

[396]   Rene von Schomberg, 'The precautionary principle and its normative challenges' in Elizabeth Fisher, Judith Jones and Rene von Schomberg (eds), *Implementing the Precautionary Principle: Perspectives and Prospects* (Edward Elgar 2006) 21.

is intricately connected to the possibilities of its comprehensive representation.[397] The assumption here is that better informatively characterized scientific uncertainty improves decision-making, requiring scientists like risk analysts whose work affects policy-making to make better reporting of the uncertainties that are associated with their findings.[398] This could diminish the masking of error or implicit cultural biases in the scientific enquiry or findings, and promotes a self-critical attitude to the scientific enquiry significant to the process.[399]

However, identification, acknowledgement and characterization of scientific incertitude that are inherent to a technological process may not be a matter that can be left exclusively to expert activity, and would need wider participation and engagement. Wynne's work regarding the effect of Sellafield nuclear plant on the neighbouring Cumbrian sheep-farming communities is extremely important here. He highlights the experience of lay contestations, assertions based on *real* everyday experiences, which have been deemed irrelevant by expert construction for decades, being recognized as scientific uncertainty at a much later stage. He effectively argues that scientific uncertainty is not necessarily exposed through expert disagreements, but rather that public contestations about scientific certainty informed by everyday experiences can lead to gradual recognition of those concerns by some experts, which may later lead to minority opinions and disagreements between experts.[400] However, this recognition may be considered too late in cases where irreversible or extremely serious harm could have occurred, a scenario which is traditionally identified as a key rationale for the invocation of the precautionary principle. In situations where the need for precaution is already felt or demanded, particularly in cases of radical incertitude like ambiguity and

---

[397] Katie Steele, 'The precautionary principle: A new approach to public decision making?' (2006) 5 *Law, Probability and Risk* 19, 23.

[398] Frances M. Lynn, 'The interplay of science and values in assessing and regulating environmental risks' (1986) 11 *Science, Technology and Human Values* 40.

[399] For instance: Les Levidow, 'Precautionary risk of Bt. Maize: What uncertainties?' (1997) 83 *Journal of Invertebrate Pathology* 113, 116. He underlines the importance of the precautionary principle during risk analyses of GMOs, whereby attention to scientific uncertainty encourages scientists to sharpen their critical approach to both the substance and methodology of their enquiry.

[400] Brian Wynne, 'May the sheep safely graze? A reflexive view of the expert–lay knowledge divide' in Scott Lash, Bronislaw Szerszynski and Brian Wynne (eds), *Risk, Environment and Modernity: Towards a New Ecology* (Sage 1996) 44.

ignorance, how can expert estimations of uncertainty be taken as the final word? While the object of tying the trigger of the principle to scientific advice may be to discipline the potential arbitrary use of the principle, the situation of radical incertitudes may be completely missed out in such scenarios. Radical incertitudes denote situations of limitations of under- standings and rationalities emanating from techno-scientific spaces, where letting the trigger of the principle be controlled through them would mean that techno-scientific spaces decide the limitations of their own understanding. Since these are exactly the instances where the regulatory innovation of precaution is most important, they could end up making legal implementations of precaution much similar to classical risk models, an aspect elaborated later in this chapter regarding the experience in EU regulation of GMOs.[401]

The earlier-mentioned collaboration provides a sympathetic and yet critical account of precaution, where it identifies three important tracks of criticism regarding the legitimacy and workability of the principle. 'Some critics describe the principle as a "no-risk" and "non-science based principle"; however, in practice regulators using the principle have paid more attention to science and have considered a wider array of possible regulatory response'.[402] A second identified track of criticism 'accuses the principle of providing no clear guidance to decision-makers, and ascribes a lack of internal coherence to the principle due to its various contrasting formulations'.[403] The collaborators remind us that such criticisms blindly ignore the nature of a legal principle (its inherent flexibility) and influence of legal culture, and thus suggest that the aforementioned authors do not know the basic methodological principles of comparative law.[404] An important (third) fear identified is that the principle will be a means of justification of arbitrary action or ulterior motives because decision-makers do not need to rely on 'facts' for making decisions.[405] This overlooks the reality that in circumstances of

---

[401]   Text to nn. 449–450.

[402]   Fisher, Jones and von Schomberg, 'Perspectives and Prospects' (n. 393) 4, citations omitted.

[403]   Ibid 5, citations omitted.

[404]   Ibid 5.

[405]   Giandomenico Majone, 'What price safety? The precautionary principle and its policy implications' (2002) 40 *Journal of Common Market Studies* 89; Gary Marchant and Kenneth Mossman, *Arbitrary and Capricious: The Precau- tionary Principle in the EU Courts* (AEI Press 2004). Connected to this fear is the attempt to 'rein in' or discipline precaution and limit it to environmental protection, excluding other realms, like, say, health protection. See Han Somsen,

scientific incertitudes, what is an arbitrary decision is itself open to question, and relying on the *facts* in such circumstances is profoundly problematic, because the nature of scientifically establishable facts is itself in doubt.[406] It is also connected to public memories of a number of abject failures of scientific inputs to protection of human health and the environment, for instance in the denial of already visible hazards of industrial asbestos, pesticides like DDT, thalidomide, tobacco and BSE, for decades in some cases, giving strength to the public demand (which also underpins the assumption of the precautionary principle) that risk governance and related regulation of technologies ought to have wider social and political input and control.[407]

The epistemic difficulties of knowability of relevant regulatory facts beforehand have forced a move (grudging in some cases) beyond the principle of prevention, which bases itself on 'provable scientific facts', towards the acceptance of precaution as a guiding norm behind regulatory action.[408] After all, it is all but obvious that in a modern welfare state any public action known to cause harm to the environment or health needs to be prevented, and the real challenge is the epistemic difficulty in being able to know.[409] As we saw earlier in Beck's Risk Society, the controversy is rather about the process of acknowledging and anticipating uncertain and unknown hazards, and constructing statistical models in this regard, requiring precautionary approaches in a world of reflexive science and other institutions. If the normative justification of the principle is acceptable, what transpires as the most difficult challenge is to decide the social ambit of its trigger. There are reasons of governance that tie the trigger to scientific/expert opinion. However, as mentioned

---

'Cloning Trojan horses: Precaution in reproductive technologies' in Karen Yeung and Roger Brownsword (eds), *Regulating Technologies* (Hart Publishing 2008) 221.

[406]   Elizabeth Fisher and Ronnie Harding, 'The precautionary principle and administrative constitutionalism: The development of frameworks for applying the precautionary principle' in Elizabeth Fisher, Judith Jones and Rene von Schomberg (eds), *Implementing the Precautionary Principle: Perspectives and Prospects* (Edward Elgar 2006) 113; Harremoës (n. 364).

[407]   Brian Wynne, *Risk Management and Hazardous Waste: Implementation and the Dialectics of Credibility* (Springer-Verlag 1987).

[408]   See generally Beck, *Ecological Politics* (n. 315) and *Risk Society* (n. 306).

[409]   'Many modern treaties exhibit this new status quo by mentioning ... solely the precautionary principle, whereas it is evidently not the parties' intention to combat uncertain threats while leaving certain threats alone', Trouwborst (n. 367) 192.

earlier, if the difficulty is also about limits of expert/scientific rationality, then tying the applicability of the principle to expert understandings and predictive capacity is suspect. Further, the wisdom of tying the trigger to the same space becomes spurious, when techno-scientific spaces that generate these technologies judge their own limitations and also profit from them.

4.1.1   Multiplicity, Vagueness and Ambivalence

Given the lack of a universally accepted definition and interpretation of the precautionary principle, including its role in policy, legislation and decision-making, analysis about the social dynamics of precaution in techno-scientific controversies is surprisingly scarce. A multiplicity of meanings – seen first in official legislative instruments, policy documents and court decisions, secondly in sympathetic criticisms to strengthen the objectives and mechanisms of precaution for greater levels of protection from potential hazards and stronger public participation, and thirdly in normative and practical objections to the implementation of the principle – all begin their analyses from drastically different characterizations of the principle.[410]

Recent sociological attention discusses the principle's topical recurrence in the Spanish and other European media, based on personal interviews with a variety of relevant social actors (both supporters and critics of GM technology).[411] An earlier qualitative survey, by the same authors, of media reports about the debate on GMOs in Spain demonstrates that the precautionary principle is commonly used by a wide variety of social actors in public and political debates about health and the environment.[412] A survey of literature by the authors discusses the role of the principle as a regulatory guideline during techno-scientific conflicts and the role of science in regulatory decision-making, including its introduction in a range of legislation in the EU. Two contrasting ideal-types of implementing the principle are identified here as informed by different visions of the principle, viz., (1) 'as one element (among

---

[410]   See for a review of literature in this regard: Carolina Moreno, Oliver Todt and Jose Luis Lujan, 'The context(s) of precaution: Ideological and instrumental appeals to the precautionary principle' (2010) 32 *Science Communication* 76, 77.

[411]   Ibid.

[412]   Jose Luis Luján and Oliver Todt, 'Precaution in public: The social perception of the role of science and values in policy making' (2007) 16 *Public Understanding of Science* 97.

many others) of a process of decision making guided basically by scientific knowledge, which understands scientific uncertainty as a temporary lack of knowledge', and (2) 'as a decision criterion to select technological trajectories, precaution being more important than scientific knowledge in taking decisions'.[413] These ideal-types are portrayed as growing from opposing ideological positions on the role of science in decision-making.

The sociological studies reflect on the two ways of understanding this fundamental role and importance of science in regulation and public policy-making: one group of social actors (who subscribe to the 'technology governability interpretation') tends to hold the view that technological progress and the application of new technologies (such as biotechnology), under normal circumstances, are always positive for society. Although the existence of scientific uncertainty is not denied, this group's interpretation is that uncertainty is a temporary lack of scientific knowledge, a knowledge that will be generated over time through strict application of expert knowledge and scientific analysis. This makes precaution subsidiary to scientific risk analysis and turns it into just one among several aspects of decision-making. In principle, and in spite of uncertainty, the undesired consequences of technology are supposed to be manageable through regulation. The other group of social actors (who subscribe to the 'technology selection interpretation') considers precaution to be an overarching decision-making criterion, especially because technological progress is seen as not only holding benefits but also risks and disadvantages for society. Here, scientific uncertainty is interpreted as intrinsic to certain technologies because of their complexity. That turns uncertainty into a form of scientific ignorance about the future effects and impacts of a new technology that is not sufficiently predictable or controllable. The precautionary principle, in this paradigm, would guide the de-selection of technologies whose benefits are unclear and whose negative impacts are possibly severe and uncontrollable. Science, in turn, is subsidiary to the debate about the reasons for accepting or rejecting the technology in question, even possibilities of its uncertainties, benefits and costs.[414]

The key difference between the two views, as delineated in this study through the survey and its literature review, is the interpretation of the role of scientific knowledge in decision-making. One group favours decisions being guided mainly by scientific knowledge, with possible

---

[413]   Moreno and others (n. 410) 82.
[414]   Ibid.

negative effects to be managed if and when they occur (ex post regulation). The other group argues for intervention in the process of innovation in order to minimize possible negative effects materializing (ex ante regulation), using the precautionary principle as a decision-making guideline (while considering scientific knowledge as merely one among several sources of information). 'In other words, it is the understanding of what role science should play in regulatory and policy decisions, which ultimately sets apart those who desire precaution to be used as a fundamental guide for policy making, from those who desire precaution to be subject to science'.[415]

Prioritization of the former (even by some who acknowledge the conceptual and normative importance of the latter for reasons of major industrial and regulatory failures) is often defended for its easy implementability and the need to have clear and consistent frameworks that classical risk-based approaches might appear to provide. Some commentators suggest that the experience of EU regulation shows that the core of the principle in Europe resides in the discretion of policy-making institutions to legislate and to make complex assignments of political, social and other considerations, in a proportional manner.[416] However, such complex assignments and considerations of proportionality are also intricately connected with grave ascriptions by some that governing elites exhort the acceptance of the principle as a guiding light, while in reality subverting, emaciating and cadaverizing the principle during its implementation, in the name of workability, proportionality and anti-discrimination. Here, the assertion appears to be that while the general applicability and normative importance of the principle are upheld by governance structures, and often used for reasons of legitimacy and trust, the reality of implementation shows its subversion, 'reduced as it were from a tiger to a domestic cat'.[417] Thus, in this account, opponents of the principle, including influential governing elites, put pressure on the principle by accepting its generality and then vigorously seek to alter the principle and emaciate it. This trend of interpretative emaciation of the principle, noted by various commentators, is intricately connected to the official view that the precautionary principle offers an unwelcome and technically insupportable alternative to decision-making predominantly informed by scientific experts. This comes from a particular

---

[415]   Ibid 77.

[416]   See for more: Ilona Cheyne, 'The precautionary principle in EC and WTO law: Searching for a Common Understanding' (2006) 8 *Environmental Law Review* 257.

[417]   Applegate (n. 366).

political position that informs the regulatory assumption that economic expansion and technological innovation is the best route to social welfare, including improved environment and human health.[418]

Without contesting the need for thinking through appropriate strategies for implementation in a proportionate, non-discriminate and non-arbitrary manner,[419] the concern for critique is to ensure that these attempts keep intact the fundamental underlying rationale of the precautionary principle, which is closely connected with the need to democratize expertise in general and scientific expertise in particular. In its core, the principle limits the pure reliance on technical/scientific expertise to assess the safety or acceptability of existing and new technologies or industrial and commercial activities, given the recognition of the epistemic and ontological limits of scientific discourses in regulation. This brings us to an important tension between the core concerns that have strengthened precaution, namely the common-sense of protecting the vulnerable from harm, even when science is unable to ascertain or even formally anticipate it. On the other hand, this identification of the limits of scientific discourses is understood in many dominant institutional spaces as an issue to be decided by experts. When risk, uncertainty and weight of evidence as characterized by scientific experts play a determinative role in the trigger, impact and extent of the applicability of the principle, the principle is seemingly linked to concrete factors. Thence, the operationalization of the principle betrays the very reason why there are calls for actions to be based on precaution in the first place, which is related to the difficulty of the classical risk framework and the evidence it produces.

Notwithstanding varying hues of the principle, the danger of technocratic governance structures merely using the principle to claim legitimacy while simultaneously subverting the principle in its application needs to be underlined. This can make the principle useless and the representative and cognitive problems that led to the invocation of the principle will continue to be left unaddressed.[420]

---

[418] Ibid.

[419] Credible proposals at alternate triggers of precaution can be seen. For instance, see Andy Stirling, Ortwin Renn and Patrick van Zwanenberg, 'A framework for the precautionary governance of food safety: Integrating science and participation in the social appraisal of risk' in Elizabeth Fisher, Judith Jones and Rene von Schomberg (eds), *Implementing the Precautionary Principle: Perspectives and Prospects* (Edward Elgar 2006).

[420] 'For instance, all three major environmental threats in the current decade have, in practice, seen cautious approaches to precaution: climate change,

4.2 REGULATORY EMPLOYMENT OF PRECAUTION IN GOVERNANCE OF GMOS

Placing an attempt to understand the employment of the principle in the EU regulatory framework of GMOs within the aforementioned broader controversial settings would further the investigation of how the system is responding to norms of trust and engagement. The second chapter explored the contours of the current regulatory framework on GMOs. It identified three important stages in the decision-making for commercial release: a risk assessment by EFSA, a national authority or both, followed by risk management through a Comitology process and possibilities of a subsequent re-examination through environmental guarantee and product safeguard measures. This section takes this identification forward by investigating the concrete differences that invocation of precaution has effectuated in these stages. The contours of differences that will be identified are twofold: do such invocations correspond to a meaningful incorporation of precaution in the regime, and does this incorporation foster trust and encourage wider public engagement.

Though this section will only examine employment of the precautionary principle directly related to mechanisms of release, it is important to remember that precaution can be implied (at the very least) in other stages of regulation like coexistence and labelling.[421] The Deliberate Release Directive expressly affirms that measures under the Directive 'to approximate laws, regulations and administrative provisions of the Member States and to protect human health and environment' ought to be in accordance with the precautionary principle.[422] The Directive requires Member States to adhere to the principle to ensure taking all appropriate

---

hazardous chemicals, and over fishing': David van der Zwaag, 'The precautionary principle and marine environmental protection: Slippery shores, rough seas and rising normative tides' (2002) 33 *Ocean Development and International Law* 165.

[421] As mentioned in Chapter 2, though Commission guidelines claim that coexistence is only about economics, there is a precautionary logic to coexistence measures and the protection of the organic chain of production. See for more, Lee, 'Governance of coexistence' (n. 225). However, the invocation of the precautionary principle is far more explicit in the legislative instruments on Labelling and Traceability.

[422] Article 1 Deliberate Release Dir. This is reiterated in its recital that 'the precautionary principle has been taken into account in the drafting of this Directive and must be taken into account when implementing it', Recital 8.

measures to avoid adverse effects on human health and the environment.[423] Further, the choice in the Directive for a case-by-case evaluation of the socio-economic, environmental and health risks for every application of release of GMOs, as opposed to the doctrine of substantial equivalence that is applied in some other realms, shows the precautionary intent in the EU legal regime.[424]

However, in the classical bifurcation of risk into technical analysis and political management used in the release mechanisms, technical assessment is placed at a prior stage in a manner pre-determinative of the whole safety inquisition. Crucially, the Commission in its Communication has negated the use of the precautionary principle at the stage of risk analysis, terming precaution a politically accepted strategy of risk management.[425] Earlier in its argument before the WTO Appellate Body in the *EC-Hormones I* dispute, the EC had asserted the need to apply the principle in both the phases of risk management and risk assessment.[426] The position of the Commission in its Communication is in contravention to the legislative injunctions in the Deliberate Release Directive that require general principles for environmental risk assessment to be in accordance with the precautionary principle and acknowledge 'delayed effects on human health or the environment that may not be observable during the period of release, but only become apparent as direct or indirect effects either at the stage or after the decision of the release'.[427] This legislative prescription is persistently violated by the influential position of the Commission that the principle is confined to risk management and inapplicable in risk assessment. Such a position of the Commission, which effectuates a restrictive application of the principle in contravention to legislative injunctions, would appear to turn precaution

---

[423] Article 4 (1) ibid.

[424] See for more on the process versus product debate: Howse and Regan (n. 133).

[425] '[W]hen there are reasonable grounds for concerns that potential hazards may affect the environment or human, animal or plant health, and when at the same time the available data preclude a detailed risk evaluation, the precautionary principle has been politically accepted as a risk management strategy in several fields': Commission Communication (n. 376).

[426] WTO Appellate Body Report, *EC Measures Concerning Meat and Meat Products* (*EC Hormones-I*), WT/DS26/AB/R, WT/DS48/AB/R, 16 January 1998, para. 16. See further, Sara Poli, 'The EU risk management of GMOs and the Commission's defence strategy in the biotech dispute: Are they inconsistent?' in Francesco Francioni and Tulli Scovazzi (eds), *Biotechnology and International Law* (Hart 2006).

[427] Annex II B.

into an adjunct to the classical risk paradigm with little real impact of its own. Whether this prima facie position is indeed correct requires further investigation in the three aforementioned stages, viz., risk assessment and risk management before release, as well as re-examination of safety considerations during environmental guarantee and product safeguard measures.

4.2.1   Central Role of EFSA in the Current Regulation of Risk

The breakdown of the earlier release mechanism for transgenic food, due to the high level of distrust between Member States regarding their risk assessment procedures,[428] was sought to be tackled by a certain degree of centralization of risk assessment processes through EFSA.[429] EFSA has important responsibilities for EU risk assessment in general, and plays a key centralizing role in GM food regulation in particular. Of the eight scientific panels in EFSA seeking to cover risk analysis in various stages of the complete food chain, one panel is entirely devoted to the analysis of GMOs.[430] The Authority is responsible for centralized 'scientific' risk assessment, through its key position in the risk ascertainment processes, and plays a palpably important role in the attempt at harmonization of the release of GMOs.[431]

There are multiple aspirations for EFSA as an institution, including serving as the primary point of scientific reference, and acting as the catalyst for scientific consensus in the EU through networking with Member States and EU scientific institutions. The White Paper on Food Safety recommended the setting up of this organization that shall be 'the

---

[428]   Amply demonstrated by the facts in the *Greenpeace* decision (n. 186).

[429]   Set up under the Food Safety Regulation (n. 192). The management board of EFSA comprises fourteen members from various Member States' organizations, one representative of the Commission, one from a consumer association, and four from agriculture and industrial associations. The board is responsible for appointing the Executive Director and members of the scientific panels, and the setting of annual budgets.

[430]   See for an informative account of the policy developments and different proposals that led to the formation of the Authority: Alberto Alemanno, 'The European Food Safety Authority at five' (2008) 1 *European Food and Feed Law Review* 1.

[431]   See Sebastian Krapohl, 'Credible commitment in non-independent regulatory agencies: A comparative analysis of the European agencies for pharmaceuticals and food stuffs' (2004), 10 *European Law Journal* 518.

scientific point of reference for the whole Union'.[432] The Authority is mandated to provide independent scientific advice and technical support for EU legislation and policies in all fields that have a direct impact on food and feed safety, as well as on related risks.[433] The institution is entrusted with assessing the risk of regulated substances, such as GMOs, following notification procedures and time schedules established by EU legislation such as the Deliberate Release Directive and the Food and Feed Regulation. It is also required to monitor specific risk factors, identify and characterize emerging risks, and to develop, promote and apply new and harmonized scientific approaches to hazard and risk assessment of food and feed.[434]

The general EU regulatory scheme aims to foster scientific consensus in the Union through EFSA in a number of ways. One of the objectives of setting up EFSA is the quest to keep the precautionary principle in check by adopting a uniform scientific basis throughout the EU,[435] and by requiring EFSA to base its risk assessment on available scientific evidence.[436] The risk managers are required to take into account risk assessment, and in particular the opinions of the Authority referred to in Article 22 of the Food Safety Regulation, other factors legitimate to the matter under consideration, and the precautionary principle.[437] Precaution is stipulated under the Food Safety Regulation to entirely reside in the management stage, and is to be applied on a provisional basis in specific circumstances. This application hinges upon a report of technical assessment of available information by EFSA where the possibility of harmful effects on health is identified and a finding that scientific uncertainty persists is recorded.[438] This general scheme, which conceptualizes the dominance of the techno-scientific space, recognizes an extremely weak role for other publics during the preparation, evaluation and revision of food law, including through consultations of the lay public either directly

---

[432] White Paper on Food Safety, COM (99) 719, 12 Jan. 2000. It sketched out the main features of the recommended institution, leading to a first proposal for the EFSA regulation: 'Proposal for a regulation of the European Parliament and of the Council laying down the general principles and requirements of food law, establishing the European Food Authority, and laying down procedures in matters of food safety' COM (2000) 716.

[433] Article 22, Food Safety Reg. (n. 192).

[434] Article 29, ibid. See for more: Alemanno (n. 430).

[435] Preamble 20 and 21 of Food Safety Reg. (n. 192).

[436] Article 22 ibid.

[437] Article 6 ibid.

[438] Article 7 ibid.

or through representative bodies.[439] Here, in situations where there are reasonable grounds to suspect that a food or feed may present a risk for human or animal health, the Authority is advised, depending on the nature, seriousness and extent of that risk, to inform the public of the nature of the risk to health, identifying the risk that it may present to the fullest extent possible, and the measures taken or about to be taken to prevent, reduce or eliminate such risk.[440]

In cases where there is an overlap between applicability of the Deliberate Release Directive and the Food and Feed Regulation, EFSA is required to consult all national competent bodies while it carries out risk assessments. Again, during the process of authorization of seeds or similar material, EFSA has to consult a national competent body, which shall carry out an environmental risk assessment.[441] EFSA as a 'scientific point of reference' is required to act 'in close cooperation with the Competent Authorities in the Members States' carrying out similar tasks to those of the Authority.[442] Importantly, the impact of such consultation processes on the ultimate decision-making is unclear; though it is without doubt that transparency of the decision-making process at this stage would make GM regulation more robust and open to high profile contestations, potentially giving competing Member State scientific positions an important role to play.[443]

Chapter 2 noted that food safety assessment procedures under the Food and Feed Regulation are far more centralized than in an environmental risk assessment under a joint application under the Regulation and the Deliberate Release Directive, and less open to public and citizen involvement. Here in cases of divergence or even potential sources of divergence with national food agencies or other similar bodies, the Authority is required to take the initiative to ensure the sharing and identification of potentially contentious scientific issues between these expert bodies.[444] In this endeavour to foster scientific consensus through the Authority, there exists significant institutional encouragement in the regulatory framework to resolve these differences internally. Although EFSA is required to circulate draft reports among national technical authorities (and in certain

---

[439]   Article 9 ibid.
[440]   Article 10 ibid.
[441]   A joint reading of Article 6 (6) of Food and Feed Reg. and Food Safety Reg. The real impact of such consultation processes on the ultimate decision-making is unclear, though it can improve transparency in decision-making.
[442]   Article 22 (7) Food Safety Reg.
[443]   A joint reading of Art. 6 (6) Food and Feed Reg. and Food Safety Reg.
[444]   Article 30 Food Safety Reg.

enumerated cases, consult them) as well as to make the report publicly accessible, the effect of these on the final decisions appears minimal.[445] Further, in extreme cases of divergence between expert bodies that cannot be resolved between these bodies, a joint document identifying the uncertainties and contentious issues is to be publicly presented.[446]

Here, the precautionary principle in GMO regulation reflects an imagination of scientific uncertainty that is temporary, which can be routinely overcome by more research, with no recognition of the systemic ambiguities and ignorance inherent in the post-normal condition of the scientific enterprise. Would EFSA's statutory mandate to be mindful of divergences of expert opinion among different Member State authorities, and to foster cooperation among them, be able to make its assessment precautionary in substance? It would appear that this would be insufficient, since precaution also anticipates scenarios at the limits of scientific rationality and expert understanding, and not mere disagreements between experts. Concerns of divergence about scientific incertitudes can only be partly tackled through attention for minority expert opinions, as various sociological works demonstrate the limitations of expert networks to recognize divergences that laypublics have already pointed out. Wynne effectively demonstrates that scientific uncertainty is not necessarily exposed through expert disagreements, but rather that public contestations about scientific certainty informed by their everyday experiences can lead to gradual recognition of those concerns by some experts, which may later lead to minority opinions and disagreements between experts.[447] The effect of restricting the precautionary principle to risk management is the prevention of public fact-gathering that can lead to the questioning of existing frames of technical assessment, often uncomfortable to dominant conventional scientific orthodoxies, which in turn can trigger new scientific research that recognizes possibilities of various kinds of scientific incertitudes, including uncertainty, ignorance and ambiguity. In previous decades, these kinds of steps could have led to earlier identification of hazards in the use of substances like tobacco, asbestos and Agent Orange, contributing significantly to earlier awareness and better protection of health and the environment.[448]

---

[445] Lee, 'EU Regulation' (n. 15) 149–150.
[446] Article 30 (3), (4) Food Safety Reg. See further, Alemanno (n. 430) 16.
[447] Wynne (n. 400).
[448] Kastenhofer (n. 356) 323–324: 'Funtovicz and Ravetz borrow the term "popular epidemiology" from Brown to refer to epidemiological investigations by non-epidemiologists. In the case of mobile communication technology, these played an important role in raising public awareness and fostering a critical

Further, allegations have been persistent that the framing of GMO risk assessment processes by EFSA has often been one dimensional, inter alia excluding cumulative and long-term environmental effects and bio-diversity issues. These have led to periodic exhortations from the Council to improve the scientific consistency and transparency of EFSA decisions on GMOs.[449] A profound contradiction is perceivable in this situation. On the one hand, the Deliberate Release Directive invokes the precautionary principle as a cornerstone of EU GMO regulation, and emphasizes that the general principles for environmental risk assessment should adhere to the precautionary principle. On the other hand, the Commission insists that the principle is inapplicable during risk assessment, and can only be triggered by a scientific evaluation by technical bodies like EFSA.

The current regulatory framework appears very much within the classical risk paradigm. It is apparent that EFSA plays a central role in the framework for release of GMOs in a number of ways. Prior to decisions for release of specific GMOs, there are few overtures for public fact-gathering that precaution employed in technical risk assessment promises to bring in. In the bifurcated risk regulation that leads to a decision on release, the technical risk assessment conducted by EFSA precedes risk management in a manner quasi-determinative of the con-tours of the whole exercise. Though the risk managers are not obliged to adhere to the technical opinion of the Authority, regulators are required to explain deviations from the relevant technical advice.[450] Placing the scientific/technical assessment exercise at such a vantage position at the beginning of the regulatory process ensures that the construction and definition of risk are strongly determined by the scientific body, despite grave public reservations about the competence and legitimacy of such an exclusive reliance on technical expertise. Thus, the palpably central role played by the technical risk-assessors in the current GMO regime, in an

---

position. They were undertaken not by absolute *laymen* but by concerned medical practitioners. Later on, these studies were severely criticized methodo-logically, charged with statistical flaws or deficient appraisals of EMF exposure. But overall, they succeeded to point at blind spots and so far neglected heuristic horizons in which to search for harmful effects. They did so by combining a critical attitude and general concerns with basic expertise in scientific research and an epistemic approach that built upon personal observations and a focus on local populations'. Citations omitted, emphasis added.

[449] For instance, Minutes of meetings of the Environmental Council, Com-mission Press release IP/06/498, 27 June 2006, 12 April 2006; 2 December 2005 and 9 March 2006; See also, 'Cracks start to show in EU GMO policy' (6 April 2006) *EurActive*.

[450] Article 7 Food and Feed Reg.

insistently non-precautionary manner,[451] makes the release procedures appear well within the classical model of risk. Since the importance of wider public fact-gathering at the stage of risk assessment to an implementation of the precautionary principle is negated in the current GMO regime, whether precaution employed during risk management of release of GMOs could involve wider public inputs needs investigation.

4.2.2   Risk Management through Committee Procedure

In the second chapter we saw that politically appointed scientific experts in the GMO regulatory Committee repeatedly failed to reach a qualified majority to accept or reject Commission recommendations to release specific GMOs for agriculture. Such deadlock has persisted in the Environment Council, and has led the Commission to permit releases of specific GMOs on the basis of the prior positive technical report from EFSA. In most cases, such permission to release was despite a simple majority of both experts (in the regulatory committees) and the Council voting against the releases. Permissions to release GM crops in the face of persistent and considerable opposition have been doubly controversial because the Commission has insisted that the precautionary principle can only be employed in risk management, and not in the context of risk assessment. Does this not then mean that in cases of such systemic deadlocks at the management stage, when permissions are solely based on the proceedings of the technical assessment, the whole GM release-structure should be considered anything but precautionary?

There is a related controversy that adds to the conundrum of the application of the principle in the GMO regime. Accompanying the Commission's insistence that precaution can only be employed in risk management is its stipulation that the principle can be triggered only through the technical characterization of scientific uncertainty by an expert body like EFSA. Further, EU courts and the Food Safety Regulation have insisted that scientific advice can be overruled by a regulator (risk manager) only through invocation of what could be considered as superior scientific evidence.[452] These two lines of reasoning mean that the trigger of precaution itself revolves around discourses of techno-scientific risk. It is usually the very same institutional arenas which generate and fine-tune these technologies that also provide the information, data and

---

[451]   See the insistence of the Commission in its Communication (n. 376).
[452]   Scott, 'Precaution in European Courts' (n. 373).

expertise upon which these scientific judgements regarding uncertainty can be made. In an age of post-normal science:

> forces behind technological development are systematically biased in the direction of generating and neglecting certain kinds of undesirable consequences, ... and where profits drive the deployment of a technology and there is no profit in identifying its harms, there is no reason to think that the technologist will spend adequate resources to identify the harms.[453]

Such convergence of interests in institutional arenas makes application of the precautionary principle problematic, since concrete and legitimate exceptions to use technical expertise must now be initiated by the very same technological spaces. Thus it becomes clear that what constitutes scientific uncertainty is a question which is both vexed in practice and in principle, and blind reliance on scientific ascertainment of scientific uncertainties is plagued by the very same conceptual problems that accompany classical risk regimes. That is to say, in situations of uncertainty, ignorance and similar incertitudes, there is neither any institutional incentive nor a reasonable possibility for technological spaces to go all out to identify possible trajectories of hazard analysis, until the public demands it. And an insistence on preliminary risk-based characterization of uncertainty to trigger precaution chills any available possibilities within a precautionary framework to address these issues. A prescription of public involvement in fact-gathering and framing of the risk assessment process is intricately connected to the promise of precaution avoiding a purely science-based regulation, at least in principle.

4.2.3   Public Determination of Social Value through Precaution

Connected is the question of political determination between the two idealized conceptual frameworks within which the precautionary principle has to operate, viz., technology governability and technology selection, mentioned earlier.[454] The second idealized vision has an overarching decision criterion as a precautionary paradigm, according to which scientific uncertainty is intrinsic to certain technological trajectories. Here, the existence of ignorance and ambiguities about future effects and impacts invalidate the idea of accurate a priori predictions

---

[453]   James E. Krier and Clayton P. Gillete, 'The un-easy case for technological optimism' (1985) 84 *Michigan Law Review* 405, 407.

[454]   Moreno and others (n. 410).

through previous models. The limits of scientific appraisals of risk in radically new technologies become stark due to the lack of empirical data, which makes the prediction of possible risk outcomes particularly limited. This is because such probabilities are constructed from existing theoretical models and assumptions, based on analogies from other older risk models.

Such inadequate empirical predictions are influenced by social, cultural and political factors that are dominant among expert communities. Experts might not only underplay incertitudes due to these normative factors, but policy-makers might also invisibilise these factors and dress policy-making in an objective and scientific garb. Often, time constraints and demands from the industry for an urgent decision for an early-mover advantage limit a comprehensive risk assessment. This might exclude identifying and examining long-term ecological effects and other lines of enquiries that might be considered in scientific spaces as hypothetical at the time.[455] For instance, most instruments of food safety focus on the impact of a single ingredient or food regarding a specific aspect of human or animal health. The complex nature of the human body and the environment as well as the uncertainty of data and methods make risk ascertainment of new technologies in this manner inherently inadequate. Given the fragmentary nature of information and uncertainty, use of science is a necessary but insufficient basis for policy decisions. In such cases, science becomes subsidiary to other sources of reason for regulating technologies, for instance political and economic justice. This gives space for the precautionary principle to be a facilitator of choices between technological trajectories, and affords conversational space for public contestations and disagreements about the development and use of specific technologies. Ideally, this second framework has the potential to mediate such contestations where issues of social value can be brought into institutionalized evaluations of safety through, for instance, deliberation with groups outside expert systems.[456]

The potential to include social values in precaution is patently negated in the EU GMO regime by the insistence that precaution can be employed only at the risk management stage through a trigger of a technical finding of scientific uncertainty. Hence an opportunity to address issues of public trust, engagement and acceptability through a precautionary approach to the regulatory tool of risk is lost. In this

---

[455] Ortwin Renn, 'Style of using scientific expertise: A comparative framework' (1995) 22 *Science and Public Policy* 147.

[456] See for a discussion on the possible regulatory connections between utility and release of GMOs, see text to nn. 824–829.

absence of overt regulatory attempts to influence the shaping of the development and employment of technology, law appears to be banking on the market to choose between technological trajectories, including through consumer choice. Such reliance on the market for incorporating social value during regulation has serious limitations, including in fostering public trust and engagement. To be sure, any institutional attempts to incorporate considerations of social value in a participatory manner, through the precautionary principle, will not be an easy endeavour, and would be beset with epistemic and other complexities that need attention and energy. While calls for incorporation of social values are often characterized as messy and unworkable, it is difficult to deny its normative advantages and cognitive superiority over traditional risk frames. The Dutch Scientific Council for Government Policy has called for regulatory attention to foster institutional innovations towards this end. It reiterated the links between the second ideal-type of precaution with the value of public participation in the regulation of the development and use of technology,[457] requiring the involvement of political norms and not merely technocratic ascertainment.[458] Thus it is clear that a meaningful employment of precaution ought to include a wider engagement and participation in risk governance. Further, it is important to emphasize the application of a version of the principle that is not 'cadaverized by ruling elites', but which is precautionary not only in name but also in substance.[459] Recent studies on deliberative exercises that sought citizen participation in issues concerning technological risk are important here.[460] An empirical focus on the aforementioned deliberative exercises within these studies concluded that 'deliberative minipublics almost always produce recommendations that reflect a world

---

[457]   One that recognizes the vulnerability of people, society and the natural environment, necessitating a proactive approach to seek uncertainties, and in various respects and essence differs from the classical risk approach: WRR (n. 2) 126–127.

[458]   Ibid 134–135: 'Precaution requires political organization ... its introduction will require the introduction of new institutional arrangements'.

[459]   Trouwborst remarked, 'action despite uncertainty, and not because of it' (n. 367) 29.

[460]   John S. Dryzek, Robert E. Goodin, Aviezer Tucker and Bernard Reber, 'Promethean elites encounter precautionary publics: The case of GM foods' (2009) 34 *Science, Technology and Human Values* 263. The study examines construction of mini-publics about GM foods in France, the US, Canada, the UK, Australia, Denmark and Switzerland.

view more precautionary[461] than the Promethean[462] outlook more common among governing elites'.[463] This is, the studies insist, 'despite varied circumstances of the establishment of these exercises by policy making elites who either commission these exercises or are the intended audience'.[464]

> Whatever the complexion of the political elite, the outcomes of the deliberative exercises are normally precautionary. If elites are promethean this poses some major question about their ability to accommodate a more deliberative democracy, at least when it comes to technology and risk. Since, ideologically, it would appear that one of the top priorities of contemporary national governments is to ensure economic competitiveness in a globalizing world, where large costs can be expected for a state due to departure from promethean facilitation of technological innovation, diffusion and marketization.[465]

In situations of precautionary public opposition to technological risk, the authors note that national governments find it hard to legitimize their commitment to technology and trade driven economic growth within transnational regimes like the WTO. They assert that the calculations of the costs of departure from their existing transnational commitments may not be as high as they are made out to be. Further, they continue that the scale of these economic imperatives being calculated as extremely high could be mere ideological constructions where states in fact do have room to manoeuvre away from this 'narrow alley'.[466] Though the commercialization of transgenic food and crops has often been characterised as an encounter that brings out the tension between Promethean

---

[461] The precautionary worldview, ascribed to substantial sections of the public, approaches the public uncertainties associated with any new technology with caution, placing substantial burden of proof on its proponents, ibid.

[462] Promethean approaches typically ascribe faith in the capacity and ingenuity of human beings to manipulate complex systems, including ecosystems, for their own advantage. Thus technological innovation and minimally regulated markets, though having undesirable side-effects, can be managed and controlled, ibid.

[463] Dryzek and others (n. 460) 264.

[464] Ibid.

[465] Ibid.

[466] See: Colin Hay and Ben Rosamond, 'Globalisation, European integration, and the discursive construction of economic imperatives' (2002) 9 *Journal of European Public Policy* 147; Paul Hirst and Graham Thompson, *Globalization in Question: The International Economy and the Possibilities of Governance* (Polity 1996).

elites and precautionary publics, the authors asserted that a collision course between the two realities can be avoided through weakening the link between economic imperatives and Promethean commitments. They felt that such a feat is achieved in north European countries 'to the degree they feature an ideology of ecological modernization'.[467] Thus ascriptions of safety are tied with economic growth through investment in techno-scientific spaces, imbibing modernist optimism. However, given the centrality of technology, and modes of living with it in contemporary risk societies, it is asserted that such an important choice between the different (precautionary) trajectories should be made in a far more democratic and participatory manner, as opposed to it being a technocratic decision. If a new social contract of innovation is a worthwhile imagery, revolving around a genuinely precautionary basis, then the time to innovate regulatory pathways for a radically different precautionary trigger is now.

In sum, the existence of a possible primary inclination of the release regime to allow releases, despite strongly articulated opposition of a majority during risk management, brings forth with renewed strength the point of democratic deficit. Notwithstanding the legislative exhortations and regulatory claims of precaution as the lynch-pin of EU regulation on transgenic food and feed, there is a fundamental ambiguity in the way the principle is given expression in the mechanisms for granting permissions for release. It would appear from the close examination here, that there is little precaution involved in the deliberate release of GMOs. Despite all the invocations of the precautionary principle in legislative documents, it appears that the classical model of risk continues to dominate the current framework in a manner where room for public participation is extremely limited. Further, can precaution, in the way it is intended to be employed at the stage of risk management, be any different from the way risk management is supposed to be conducted in a classical risk paradigm that does not claim itself to be precautionary? In other words, risk management in any case is envisaged to apply and arrange the technical findings in a manner that is normatively appropriate for the political community. Hence, what are the real implications of precaution only allowed at the management stage, and how different is the current GMO framework from any classical risk paradigm? From the discussion about the lack of precaution in risk assessment, and an examination of the pattern of

---

[467] Dryzekand others (n. 460) 263. They use the instance of Denmark. See, for a useful examination of public deliberations in Denmark, Maja Horst and Alan Irwin, 'Nations at ease with radical knowledge: On consensus, consensing and false consensusness' (2010) 40 *Social Studies of Science* 105.

decisions in the risk management stage, it appears that this framework is precautionary only in name. It is not even governance by committees and Comitology (which are in any case beset with the problems of democratic deficit), but it is a practice of technocratic governance by the Commission with the justification for the decision entirely based on a technical risk assessment that it expressly insists to be non-precautionary. The room for participation by the public is negligible given neither space for public fact-gathering in risk assessment, nor any effective oversight of normative and political factors that frame it. Thus, in the current EU regime, there is neither attempt at public participation, nor is there any way to represent concerns emanating from groups outside of techno-scientific communities.

4.2.4 'Emergency Responses' and Interpretative Role of EU Courts

An examination of the nature of ascriptions of safety at the stage of environmental guarantee and safeguard measures demonstrates two aspects of precaution. First, the imagination of the principle arising from EU court decisions reflects an expert-led model of the classical risk paradigm. Secondly, it also brings further clarity to the inadequacies in the employment of the principle in the regulatory framework for GMOs. As described in Chapter 2, continuation of concerns of safety is expressed in two stages of the regulation subsequent to deliberate release decisions, viz., the general environmental guarantee provisions in the TFEU and the product-specific safeguard measures under EU legislation. Though legislation providing for these safety measures may textually appear to grant scope for divergences at the national and regional levels, with possibilities of nuanced approaches that avoid classical risk-related standards of evidence, EU courts have limited such discretion. While decisions by EU courts facilitated an extension of the scope of the principle from strictly environmental issues to other realms like health regulation and consumer rights,[468] the conceptualization of the principle within these decisions appears to be well within a narrow imagination of precaution. Thus uncertainties are viewed as something to be overcome by way of further research (and not as intrinsic to the development of techno-science), where the identification of uncertainties that triggers the principle is left exclusively to experts.

---

[468]   Joined Cases T-74/00, T-75/00, T-76/00, T-83/00, T-84/00, T-85/00, T-132/00, T-137/00, T-141/00 *Artegodan GmbH* v *Commission* [2002] OJ L 31/1, later confirmed by Case T-147/00 *Laboratoires Servier* v *Commission* [2003] ECR PT 52 II-00085.

This is amply clear in the manner of interpretation by EU courts of the notion of 'scientific evidence' that can legitimately trigger an environmental guarantee. The milieu of *Austria* v *Commission* duly testifies to the conservative construction by EU courts, where the requirements of newness of evidence and specificity are expected to be demonstrated in a strict manner.[469] Such expectation is stated to be in the interests of uniformity and aspirations of the internal market, over the objectives of higher levels of protection of the environment and the health of citizens. The Court's approach of legitimizing the current regulatory measures by asking for strict scientific proof that is based on studies conducted after the original decision was taken is criticised by Fleurke.[470] She elaborates on how this interpretative requirement has rendered the guarantee provision practically useless, and that a more appropriate construction would have been 'to accept all relevant knowledge, including information that was already available at the time of adoption of the Community measure, but which for some reason has been overlooked or deemed irrelevant at the time'.[471]

The manner in which the text of the safeguard clause in the current Deliberate Release Directive ties its invocation to a positivist requirement of narrow scientific proof was noted in Chapter 2. The earlier Directive of 1990, however, provided for a safeguard clause that empowered provisional derogation by Member States, if there were 'justifiable reasons' to consider the constitution of risk to human health or the environment.[472] Though the safeguard measures from the previous Directive could be considered textually wider than in the current Deliberate Release Directive, this difference appears to have had little consequence due to the interpretation of the earlier guarantee by EU courts. In the *Greenpeace* decision, signatures of one-fifth of the population of Austria in support of the safeguard measures as constitutive of addressing possible risk to their health were adduced. Justifications based on such public concerns about expert constructions of risk, however, were not

---

[469]   Text to n. 213. See for a long list of critiques n. 216.

[470]   Fleurke (n. 215). See also, the opinion of Advocate General Sharpston in *Austria* v *Commission*: 'I do not think it necessary to look further in order to reach the view that new conclusions drawn from existing data may constitute new scientific evidence within the meaning of Article 95 (5) EC Treaty', para. 124.

[471]   Ibid (Fleurke).

[472]   Article 16 of the Directive 90/220/EEC of the Council on the Deliberate Release into the environment of genetically modified organisms [1990] OJ L 117/15.

regarded by the ECJ as 'justifiable reason' for maintaining safeguard measures.[473]

More recently, the CJEU in *Monsanto (France)* emphasized the reference to a 'significant risk which *clearly* jeopardises human health, animal health or the environment' for the application of the safeguard measure in the Food and Feed Regulation.[474] The Court read an additional requirement for the successful application of the safeguards, that the risk 'must be established on the basis of new evidence based on reliable scientific data'. This is a requirement that is not explicit in the letter of the legislation. The Court reiterated that the protective measures 'cannot validly be based on a purely hypothetical approach to risk, found on mere assumption which has not yet been scientifically verified', and may be adopted only if 'they are based on a risk assessment which is as complete as possible', 'notwithstanding their temporary character and even if they are preventive in nature'.[475] If safeguard measures are to be seen as giving specific expression to the precautionary principle,[476] it is perplexing how evidentiary demands on these measures can be as high as the general provisions connected to harmonization of a product.[477] In these aforementioned decisions, textually conceivable alternative interpretations could have avoided an insistence on strict scientific proof, as opposed to, say, a reliance on either 'other relevant factors' or scientific information that may not amount to scientific proof. It would have been more appropriate to tackle controversies about safety in a new paradigm that is more attuned to the normative claims and aspirations of precaution. These factors, combined with the Commission's controversial attempts at outlawing the Austrian safeguard bans, described in Chapter 2, clearly show the lack of an overtly precautionary approach in the safeguard or guarantee measures. This is since substantive requirements of EU courts insist on scientific proof of risk in a way that is not particularly resonating precautionary emphases.[478]

---

473   *Greenpeace* (n. 186).
474   *Monsanto (France)* (n. 195) [70].
475   Ibid [77].
476   *Monsanto (Italy)* (n. 191) [110].
477   See for a further elaboration: Lee 'Governance and Decision Making' (n. 254) 13.
478   See for a similar evaluation regarding the employment of precaution in the regulation of mobile technologies in Australia, Jacqueline Peel, 'When (scientific) rationality rules: (Mis)Application of the precautionary principle in Australian mobile phone tower cases' (2007) 19 *Journal of Environmental Law* 103.

Further, the interpretative role played by EU courts in the conservative construction of environmental guarantees and safeguard measures is similar to its track record in a plethora of other decisions that interpret the use of the precautionary principle in EU regulatory regimes. A number of important commentaries have identified this pattern in judgments of EU courts from *Danish food* to *Artegoden* and *Pfizer*.[479] Scott focuses on the stipulation in the *Pfizer* decision by the European Court of Justice by stating that:

> A regulatory measure ought not to be based on a purely hypothetical approach to risk, which has not been scientifically verified (para. 146). Thus the precautionary principle can apply only to situations where there is a risk, which although not founded on mere hypotheses that have not been scientifically confirmed, has not yet been fully demonstrated (para. 172). Secondly, within wider procedural criteria aiming at risk assessment based on excellence, transparency and independence (para. 146), while decision-makers may for reasons of democratic legitimacy disregard scientific advice such explanation may operate at a scientific level at least commensurable with the original report, of EFSA.[480]

Pertinently, Scott pointed out that the ECJ's insistence that science can only be tackled with science happens to coincide with decisions of the WTO tribunals on the point.[481] She discusses the interpretative fidelity of EU courts with the WTO adjudicators, in particular the decision of the Appellate Body in *EC-Hormones I*.[482] Though details of these WTO tribunal decisions are examined in the next chapter, it is important to note here the possible self-imagination of EU courts in their interpretative role, to fit the implementation of the principle in the EU with the scheme of the WTO framework. Scott clearly demonstrates the descriptive and normative inadequacies of the traditional conception of the role of the judiciary, as mere norm elaborators and enforcers, as opposed to the significant role wielded by courts as catalysts, in areas of normative uncertainty or complexity. In such situations, she argues, courts can

---

[479]   See Wybe Douma, 'Fleshing out the precautionary principle by the Court of First Instance – Analysis' (2003) 15 *Journal of Environmental Law* 372, 394. See further: T-13/99 *Pfizer Animal Health SA v Council of European Union*, [2002] ECR II- 3305; C-180/96 *United Kingdom v Commission* [1998] ECR I-3906 (BSE); Case T-369/03 *Arizona Chemical Bv and Ors v Commission* [2005] *ECR* II-5839; C-132/03 *Ministero della Salute v Codacons* [2005] ECR I-3465; *Monsanto* (n. 191).

[480]   Scott, 'Precaution in European Courts' (n. 373) 54.

[481]   Ibid.

[482]   Scott and Vos (n. 383) 253. *EC-Hormones I* (n. 426).

undertake a normatively motivated enquiry facilitating the realization of process values and legitimacy principles by the institutional actors responsible for norm elaboration within new governance.[483] Here the role of EU courts is crucial in unfolding a particular imagination of the precautionary principle as the only viable one for the EU, where there is a dynamic and reciprocal relationship between courts and other institutions of new governance like the Commission. An insistence on the strictly 'scientific' by the Court, including through the rejection of the hypothetical risk, say in *Pfizer* or *Monsanto (France)*, is in line with the Commission's general approach on the implementation of the precautionary principle, of strictly tying it to the cultural rigours of the scientific establishment. Tying the trigger of the principle to a scientific recommendation or proof is clearly driven by the objective to make it visibly non-discriminatory, and as an easy way of adjudicating and making precautionary measures accountable to easy judicial scrutiny and review.

These cases show that EU courts have consistently approached the employment of the precautionary principle as well within the first ideal-type (precaution within a technology governance framework),[484] since employing conventional risk assessment as a trigger for precaution is convenient and appealing for courts that are beset with complex technical arguments. The insistence of the Commission, through its Communication, that the principle of precaution can only be triggered through a scientific evaluation by a technical body converges with the general approach of EU courts here. This takes us back to the very same conceptual problems encountered with a complete reliance on techno-scientific risk, within the classical risk paradigm. Prima facie a benign step to implement the principle, this brings back the largely discredited central reliance on the expert-centric probabilistic techniques for regulation of new and controversial techniques through the back door. In effect, this goes beyond monopolizing the responsibility to identify possible uncertainties by the techno-scientific community, which is problematic in itself. Further, it also lends the same spaces that generate these risk-based probabilistic techniques the gate-keeping function to trigger precaution, in the name of a preliminary evaluation of the probability and seriousness of uncertainties.

Even within the classical bifurcation of risk, political factors were sought to be included during risk management. In contrast, every

---

[483]   Joanne Scott and Susan Sturm, 'Courts as catalysts: Rethinking the judicial role in new governance' (2006) 13 *Columbia Journal of European Law* 565.

[484]   Moreno and others (n. 410).

decision to release a GM crop in the current regime has been taken in sole reliance on the technical reports of EFSA, given the inability of the opposition (to specific releases) to increase their simple majority to a qualified majority.[485] Though the precautionary invocation in the current regime implied the recognition of the pitfalls of a classical risk framework, the sole reliance on EFSA reports for release, in contrast, is a regression. Through the invocation of the precautionary principle, the current regime appeared to recognize various conceptual and normative limitations of representation in the classical risk paradigm. An emphasis on precaution was intended to overcome the limitations of exclusive reliance on scientific rationality, which is signified in widespread public disaffection about GMOs, as well as simmering tensions and periodic rebellion from various Member States about decisions to release specific GMOs. In this supposedly precautionary regime for the release of GMOs, there was an ambition to marry scientific and social rationalities through recognizing a protracted role for political risk management. However, the embodiment of the principle in the GMO regime appears to be in name only, from the above discussion, and differences between the classical risk framework and the supposed precautionary framework are marginal. The close examination earlier in this chapter shows that fundamental opportunities to involve wider publics, including in issues of social value of technology, are missed due to the way the precautionary principle is applied in EU GMO regulation.

4.3 WHAT NOW FOR DEMOCRATIC ATTEMPTS FOR EFFECTIVE ASCRIPTIONS OF SAFETY?

The exclusion of a precautionary approach in risk assessment gives the supposed shift from the classical risk paradigm little meaning. Employment of the precautionary principle here can give a fillip to public fact-gathering in risk assessment, given the acute inadequacies faced by expert systems on risk identification and characterization, in situations of systemic incertitude where uncertainties, ambiguity and ignorance are inherent to techno-scientific processes. Rip points out how prudence leads nanotechnologists and government actors to consider social and ethical aspects at an early stage of development, 'a precaution to ensure continued progress of a nanotechnology turns into (some precaution)

---

[485] Text to nn. 161–167. The decisions can be found in the Eurlex database, and the post-Lisbon decisions on authorizations on the GMO register, <http://ec.europa.eu/food/dyna/gm_register/index_en.cfm>, last accessed 20 June 2013.

applied to nanotechnology'.[486] He pertinently argues that precaution is not merely about uncertainty, or about the possible adverse effects (and the accompanying cause for concern, appropriate proportionality or extent of measures), but also 'about addressing ignorance, and explor-[ing] the future with the help of science fiction and social science fiction'.[487] At what point one can say that reasonable grounds (in other words, scientifically informed) for regulatory concern come to exist is a tricky question that cannot be left to experts alone. Involving various sections of publics, to identify the nature and extent of uncertainties, can only sharpen and enrich the vigour of technical enquiries, and open new lines of enquiry for technical investigation, characterization and assessment.

Though blue-prints for such institutional arrangements need far more attention, an a priori position of ruling out such attempts as not workable is unsatisfactory, and belies the crucial necessity of these regulatory innovations. These engagements cannot, of course, be confused with derivatives of risk communication, or spurious attempts at engineering legitimacy in the name of participation without a genuine conversation about the range and depth of concerns that fuel contestations and disagreements. Rip's comment about nanotechnology is relevant here:

> The notion of responsible innovation is important, but should not be seen as an attempt to create harmony. The situation is one of struggle and of force being exerted to push one or the other position. Deliberative approaches can help to articulate positions and support interaction, but will not create a consensus. In other words, struggle is unavoidable, but through the contestation some learning can occur, and in this struggle, genuine invocation of a precautionary approach shifts the balance of forces.[488]

A number of proposals to make classical risk paradigms more in tune with contemporary understandings and realities, by attempting institutional changes to make precaution and public participation intrinsic in ascriptions of safety, already exist.[489] Irrespective of the viabilities of individual proposals, it is important to recognize innovations in the phase of risk assessment to widen it to non-expert communities as crucial to address the limitations of classical representation. Given the limited room for public participation that is found in the EU regulation of GMOs, this recognition is fundamental for any attempt for law to insist on rights of

---

[486] Rip, (n. 23) 281.
[487] Ibid 277.
[488] Ibid 275.
[489] WRR (n. 2) and Stirling and others (n. 419).

lay publics to participate and be represented in such regulation. The objective here is twofold. First, in situations of post-normal science effective measures of risk assessment need wider public input.[490] Rip describes how nanotechnologists, when discussing possible impacts and concerns about the technology, recognize the need to avoid at all costs the impasse that besets green biotechnology, by taking on board the societal concerns and ethical, legal and social aspects.[491] Secondly, new ways of accommodating citizens' normative preferences and political participation in risk assessment are important for a more democratic and legitimate model of risk governance. These different points necessitate bringing together a range of concerns identified by groups inside and outside of techno-scientific expert communities through new institutional arrangements of public participation.

This position implies that the modern assumption of trust invested in experts by political communities cannot be absolute in situations of post-normal science, and has to be exercised through engagement with other sections of society. Mouffe argues for the pursuit of active trust in post-traditional contexts through reflexive institutions, where expert positions are open to critique and guidance by citizens, since assumptions of passive trust have proved to be inadequate.[492] We have already seen how a monopolistic use of scientific trust in the regulation of technology is beset with a number of concerns. These include issues of institutional legitimacy (since the same space generates technologies and sits in judgement on the safety of them), issues of erosion of trust (political, economic and safety scandals that accompany the increase in control and domination of techno-science by profit) and issues of institutional capability (the logical inability of an expert system to demarcate the limits of its own rational understanding).

Though generation of active trust through interaction and engagement with various sections of society is necessary for the democratization of technology, it is a difficult task that requires considerable rethinking of existing regulatory frameworks. The normative and institutional contours of transforming expert systems into democratic public spheres need further reflection. Mouffe, for instance, insists that this reflection is an intensely political endeavour, which itself needs wider debate and discussion.[493] This can be done only by wider political engagement that brings fundamental disagreements out into the open, with a recognition that risk

---

[490] Text to nn. 76–83, 355–358.
[491] Rip (n. 23) 272.
[492] Mouffe (n. 314) 44–45.
[493] Ibid 51.

is never just about what experts ascribe as the best possible statistical probability. Existing global rules, especially in the WTO regime, may impact the contours of attempts at these kinds of regulatory innovations in the EU regulation of technologies like GMOs. Hence the next chapter describes the relevant global rules to identify the legal permissibility of such improvisation, which includes allowing wider communities at the regulatory table for a more public deliberation.

5. Participation in safety and public morals regulation: ambit in global rules

This chapter seeks to understand this effect on EU regulation of GMOs, and in particular how trade regimes (like the WTO) and environmental instruments (like the Convention on Biodiversity) provide room for public participation in the regulation of GMOs. The discussion in the previous chapters, related to the centrality of public representation and wider participation in the classical regulatory tool of risk and the principle of precaution, is continued here. An appropriate effectuation of the precautionary principle includes new institutional arrangements for public engagement in key areas of risk analysis and management, since the current disposition leaves insignificant room for public participation in safety ascriptions. An examination of global rules in GATT/WTO and the Cartagena Protocol is important in order to understand the permissible ambit for public participation, and non-expert representation in the regulatory pursuit of safety in the EU. The legal analysis of WTO law, which is a large and distinct field in itself, is restricted to this narrow enquiry towards this end. In a similar vein, the chapter also examines constraints imposed by the WTO/GATT in the protection of public morals and values on EU regulation of GMOs, since EU rule-making is supposed to proceed subject to these global rules, standards and principles.

Currently the global complex of rules on safety in the use and transport of GMOs consists of trade rules within the WTO regime, a voluntary expert health standard-setting mechanism in the Codex Alimentarius Commission that is intricately linked to the WTO regime, and environmental rules on trans-boundary movement of specific types of GMOs in the Cartagena Protocol to the Convention on Biodiversity. How these regimes interact with each other, and how they affect EU rule-making that responds to public concerns, are discussed in subsequent sections of the chapter. The next section examines the rules on trans-boundary movement of GMOs under the Cartagena Protocol. The subsequent section introduces the general framework of the WTO regime and identifies the various applicable international agreements in the regime. It notes the role of the Dispute Settlement Body (DSB) and expert committees in it. It further identifies the specific provisions in GATT, the Sanitary and Phytosanitary (SPS) Agreement and Technical Barriers to Trade (TBT) Agreement that are relevant to Members' protection of health and the environment, and examines how they are judicially interpreted to be in consonance with WTO goals of free trade.

The WTO panel decision in *EC-Biotech* is an important element in understanding the ambit of this configuration between concerns of free trade and safety, in the specific context of claims of scientific uncertainty and lack of representation and public trust in expert estimations of risk. This decision is critically examined against the background of related cases that interpret relevant provisions of the SPS Agreement. This section seeks to identify the representational inadequacies of ignoring intrinsic scientific incertitudes, which require innovation of legal and institutional mechanisms within the classical risk paradigm. This identification is done through examining three elements in the decision, viz., the question of which scientific studies are recognized as a risk assessment, how the panel treats the term 'insufficiency of scientific evidence', and the applicability of the precautionary principle. The latter directly relates to issues of fragmentation of international law, and in particular, concerns about interpretations of trade rules in the WTO regime making multi-lateral environmental agreements like the Cartagena Protocol obsolete. In the subsequent section, the functioning and normative role of expert committees in the trade regime (such as the Codex Alimentarius Commission) to the representational question is examined. It identifies the promise and limitations of Committee procedures in situations of disagreements and scientific controversies. It also underlines the limited role of regulatory committees in issues of fragmentation in international law, if the attempt is only to construct consensus within expert communities without reference to accompanying normative and political conflicts that fuel public contestations about specific technologies. This is followed by a summing up of the room available for public participation in the regulatory pursuit of safety for Member States. In the final section, we examine the existing room for protection of public values that might accommodate public concerns about GMOs within the WTO regime.

5.1   IMPACT OF THE BIOSAFETY PROTOCOL

The most recent addition in the complex of global rules on GMOs regards the transnational movement of certain types of GMOs through the Cartagena Biosafety Protocol to the Convention on Biodiversity.[494] Emphasis on a precautionary approach is discernible in the Preamble, as well as in a number of substantive provisions, which are discussed later in this section. These provisions suggest that recognizing the lack of scientific certainty due to insufficient relevant information and knowledge, regarding the extent of potential adverse effects, need not deter members in the pursuit of their regulatory objects in biosafety.[495] The Protocol recognizes that different levels of protection may be found appropriate by different Member States. These levels of protection pertain to baselines that identify adverse effects on human health as well as on the conservation and sustainable use of biological diversity due to the trans-boundary transfer, handling and use of these organisms.[496]

Primarily, the Protocol establishes a duty on states exporting specific types of GMOs that falls within the purview of the Protocol to elicit Advanced Informed Consent from the state of intended import.[497] The GMOs which fall within the purview of the Protocol, designated as living modified organisms (LMOs), are categorized into two, viz., first organisms that are intended only for direct use as food or feed, and, secondly, organisms which are intended to be introduced into the environment as living organisms, say as seeds in agriculture.[498] The obligations to elicit Advanced Informed Consent differ between these categories. Member

---

[494]   Cartagena Protocol on Biosafety to the Convention on Biological Diversity, 2000 39 *ILM* 1027 (*2000*) (referred here as the Cartagena Protocol). The Cartagena Protocol came into force on 11 September 2003 after fifty states had ratified it; three years after it was finalized and signed in Montreal on 29 January 2000. Currently (March 2012) 163 states have ratified it, including the EU and its Member States, Nigeria, Brazil, India, South Africa, Switzerland, Iran, Mexico, New Zealand, China and Japan. Prominent states that have not ratified the treaty also happen to be complainants in the WTO *EC-Biotech* dispute, viz., the USA, Argentina and Canada, as well as Australia. See for the text, and details on ratification: http://bch.cbd.int last accessed 30 March 2012.

[495]   Ruth Mackenzie, Françoise Burhenne-Guilmin, Antonio La Viña and Jacob Werksman, *An Explanatory Guide to the Cartagena Protocol* (IUCN Environmental Policy and Law paper No. 46 2003) paras 61–67.

[496]   Article 4 and Annex III.4 of the Cartagena Protocol.

[497]   Articles 8–10 ibid.

[498]   Article 3 (g) and (h) ibid: any biological entity ('capable of transferring or replicating genetic material, including sterile organisms, viruses and viroids') 'that possesses a novel combination of genetic material obtained through the use of modern biotechnology'. Article 5 (ibid) excludes the applicability of LMOs that are pharmaceuticals as these are addressed by other relevant international agreements.

States that intend to export LMOs also have labelling obligations that differ between these two categories.

Prerequisites for seeking an Advance Informed Consent for import intended to introduce LMOs into the environment include a risk analysis carried out by the intended exporter. This analysis is to be made available to the state of intended import along with various enumerated information about the LMO.[499] Though this state is free to accept or reject the application for import of the LMO, a decision to reject such imports is required to be based on a risk assessment conducted in a 'scientifically sound manner', and based on 'available scientific evidence in order to identify and evaluate the possible adverse effects of living modified organisms' to human health as well as conservation and sustainable use of biological diversity.[500] It is salient that this textual emphasis on sound science, often employed in classical risk paradigms, is accompanied by a version of the precautionary principle in the Protocol. This version of the principle emphasizes the maintenance of appropriate regulatory action despite the lack of scientific certainty arising from insufficient scientific information and relevant knowledge regarding the extent of potential adverse effects.[501] The reader might recall the discussion in Chapter 4, in principle arguing that characterization of scientific uncertainty cannot be controlled by techno-scientific communities alone for an appropriate precautionary emphasis in safety paradigms, and should involve wider public inputs and participation.[502] The contradictions between the precautionary emphasis and insistence on sound science in the Protocol are continued later in the section. It is significant that socio-economic factors can be considered for imports of LMOs in a limited category, viz., when those imports may impact the conservation and sustainable use of biological diversity.[503]

The obligations of seeking Advance Informed Consent are considered lower for LMOs that are intended only for direct use as food, feed or

---

[499] Article 8.1 and Annex I (i) ibid.

[500] Article 15.1 ibid.

[501] Article 10.6 ibid.

[502] See for the conceptual differences between a framework based on 'sound science' and the use of scientific information in a precautionary framework: Anne Ingeborg Myhr and Terje Traavik, 'The precautionary principle: Scientific uncertainty and omitted research in the context of GMO use and release' (2002) 15 *Journal of Environmental Ethics* 73, 76–77.

[503] Article 26.1 Cartagena Protocol. Here, the evaluation ought to be consistent with their international obligations, in particular with regard to the value of biological diversity to indigenous and local communities.

processed food, and are not intended to be introduced in the environment as living organisms.[504] Parties making a decision in connection with the domestic use of such an LMO are expected to offer detailed information of the decision, through the requisite procedure in a Biosafety Clearing House, including a risk assessment as well as characteristics and suggested methods of handling and use for other parties.[505] Further, there is an obligation to indicate relevant information on the label of the produce that it 'may include LMOs' and that there is no intention to introduce these LMOs into the environment of the importing country.[506] The latter information is required on the label so that they can be availed of a lower level of obligations under the Protocol. Freedom of other parties to use their own domestic regulatory framework to reach a different decision on the specific LMO is recognized, albeit with the explicit mention that those restrictions should be 'consistent with the objectives of this protocol'.[507]

In particular, the right to invoke the precautionary principle in decisions regarding import of this category of LMOs is explicitly recognized. Thus lack of scientific certainty due to insufficient scientific knowledge relevant to the extent of the potential adverse effects of the import does not prevent the party from taking any particular regulatory decision.[508] It could be argued that the textual versions of the precautionary principle in the Cartagena Protocol, seen in Article 11.8 and Article 10.6, are relatively strong versions.[509] This is because they provide for the restriction of movement of LMOs into or through the territory of an importing state, provided a lack of scientific certainty arising from insufficient scientific information and relevant knowledge regarding the extent of the potential adverse effects can be established. There is no emphasis on these measures being provisional, and insufficiency of scientific information is enough to establish a lack of scientific certainty.

Though commentators identify the recognition of the problems with the regulatory domination by techno-scientific spaces in the spirit of the

---

[504]   Article 7.1 and 7.2 ibid.

[505]   Article 11.1 and Annex III ibid. Developing countries/transition economies that lack a domestic regulatory framework may declare through the clearinghouse mechanism that they will make their decision regarding the first import of an LMO for direct use as food, feed or processing according to an Annex III risk assessment within a specified period of time.

[506]   Article 18.2 (a) ibid.

[507]   Article 11.4 ibid.

[508]   Article 11.8 ibid.

[509]   See for different versions of the principle, text to nn. 410–418.

Protocol,[510] it remains to be seen how the Cartagena Protocol will be implemented. Given the textual emphasis on 'a scientifically sound manner based on available scientific evidence',[511] the responsibility of characterization of scientific uncertainty could still be limited to techno-scientific spaces through expert committees. Generally, control of the trigger of the precautionary principle within expert communities gives techno-scientific reason and classical risk parameters a definitive say in ascriptions for safety. This is since the contours and possibilities of lack of scientific certainty will still be controlled by spaces that develop these technologies in the first place.[512] If so, it appears that the version of the precautionary principle employed in the Cartagena Protocol may be inadequate to recognize and represent public contestations about inherent scientific incertitudes. However, it is significant that the Protocol requires Members to promote effective access to information, facilitate public awareness including information about means of such access, and facilitate education and participation regarding the objectives of the Protocol.[513] Further, the Protocol exhorts Members to consult the public in the decision-making process regarding LMOs, and make these results available to the public.[514]

The ratification of supplementary rules that provide mechanisms for implementation, liability, redress and dispute resolution of the provisions of the Protocol is still pending.[515] The lack of dispute resolution

---

[510] Mackenzie and others (n. 495) 13–14. 'Even with proper risk assessment, some uncertainty may still remain and that in such circumstances countries should have the right to adopt precautionary measures to protect biodiversity and human health': para. 65.

[511] Article 15.1 Cartagena Protocol.

[512] Lee remarks that the approach to the precautionary principle in the Cartagena Protocol 'is very similar to the SPS Agreement', *EU Regulation* (n. 15) 227. The version of precaution in the SPS Agreement is generally considered as a weak version of the principle, Applegate (n. 366).

[513] Article 23(1) Cartagena Protocol.

[514] Article 23(2) ibid.

[515] Article 27 ibid. The Nagoya–Kuala Lumpur Supplementary Protocol on Liability and Redress to the Cartagena Protocol on Biosafety was opened for signature in New York on 7 March 2011, and shall enter into force ninety days after deposit of the fortieth instrument of ratification (Article 18). Fifty-four states including a number of states from the EU, Colombia, Tunisia and Panama have currently signed it, and fourteen states have ratified it till date, including Mongolia, Mexico, the EU and a number of its Member States: <http://bch.cbd.int/protocol/parties/#tab=1> last accessed 4 June 2013. The Protocol requires Parties to develop civil liability laws (strict or fault-based, as appropriate) for ascertaining damages where causation can be established (Article 4), with the

mechanisms becomes particularly important with the aforementioned textual ambiguities in the Protocol since there are multiple routes for Members to implement the precautionary principle.[516] Further, a number of conflicts with the provisions of the WTO regime are apparent, as discussed later in this chapter.[517] Potentially, a dispute resolution mechanism set up by the Protocol may have given a set of viable alternative interpretative possibilities in competition with the WTO tribunals. Before we examine the WTO system, and possible conflicts with the Protocol, it is important to emphasize the textual ambiguities and contradictions in the Protocol, as well as their effects on the nature of the precautionary principle employed in the Protocol.

Considerable contradictions in the text of the Protocol are visible in the Preambular clauses themselves, which start by

> *Recognizing* that trade and environment agreements should be mutually supportive with a view to achieving sustainable development,
> *Emphasising* that this protocol shall not be interpreted as implying a change in the rights and obligations of a Party under any existing international agreements,
> *Understanding* that the above recital is not intended to subordinate this Protocol to other international agreements ...

---

exemptions of *force majeure* and acts of war (Article 6), specifying response measures for ascertaining liability (Article 5) accompanied by time and financial limits (Articles 7 and 8 respectively).

[516]   Commentators underline the effect of tenuous negotiations and compromises involved in the drafting of the text on the considerable textual ambiguity between provisions. These include strong positions of the developing countries, and the Organization of African countries and the European communities, on the one hand, as well as particularly vocal positions of observer countries loosely called the Miami group (US, Canada, Australia and Argentina), which have not signed the document. See for more details on the various negotiating blocks of countries: Paul Hagen and John Weiner, 'The Cartegena Protocol on Biosafety: New roles for international trade in living modified organisms' (2000) 12 *Georgetown International Environmental Law Review* 697, 710; Stanley W. Burgiel, 'The Cartegena Protocol on Biosafety: Taking the steps from negotiation to implementation' (2002) 11 *Review of European Community and International Environmental Law* 53, 55–56.

[517]   See Sabrina Safrin, 'Treaties in collision? The Biosafety Protocol and the WTO Agreements' (2002) 96 *American Journal of International Law* 606. Further, for the contradictions and inconsistencies within the Protocol, and inter sethe WTO regime, see: Gilbert R. Winham, 'International regime conflict in trade and environment: The Biosafety Protocol and the WTO' (2003) 2 *World Trade Review* 131, 143.

Halfon brings out the conceptual difficulties in such broad and contradictory wordings in the Protocol, 'where the second clause, a classic savings clause suggests deference to free trade agreements, but the third seemingly reverses that'; while the first hints towards a legal approach to balance trade and environment through use of the term 'sustainable development'.[518]

Given the textual ambiguities, the construction of the precautionary principle and characterization of scientific uncertainty can be tied to 'available scientific knowledge' of other countries.[519] This could be against the spirit of precaution since these characterizations will be based on risk frames specific to, first, the concerns of different biosystems and ecological necessities,[520] and, secondly, a different level of protection in another country. It needs to be seen how notions of availability and different levels of protection are read together by a judicial tribunal. A later section in the chapter identifies the problems posed by the relationship between these two terms in the WTO framework ('levels of protection' and available in 'available scientific knowledge') and the limitations in representation of classical risk regimes. As seen in that section, scientific evidence found adequate for a lower level of protection may be inappropriate and inadequate for a higher level of protection. This can contravene a precautionary approach and higher level of protection that may be demanded by some publics in that Member State. As discussed in Chapter 4, a precautionary approach may also require the expression and consideration of a range of public concerns, wider than that recognized by expert communities. This embodiment of the precautionary principle in the Protocol by Member States depends partly on how they are constructed in those states, and partly on how the permissibilities are shaped in the interpretation of the Protocol by postnational judicial bodies.[521] Constructions of scientific uncertainty in

---

[518] Saul Halfon, 'Confronting the WTO: Intervention strategies in GMO adjudication' (2010) 35 *Science, Technology and Human Values* 307, 319.

[519] Article 15.1 Cartagena Protocol.

[520] See Annex III 9 (h) ibid regarding points to consider related to environmental risk assessment: 'information on the location, geographical, climatic and ecological characteristics'.

[521] However, whether postnational environmental tribunals will prefer the rationality and logic of stronger versions of precaution over classical risk parameters is a matter of mere speculation since such tribunals have not been established. See Stephen Hockman, 'An International Court for the Environment' (2008) 11 *Environmental Law Review* 1.

the regulation of GMOs were required to be consistent with the stipulations of insufficiency of scientific evidence under the WTO regime in the *EC-Biotech* decision. Due to a particular interpretative technique in the decision, the Protocol was found to be inapplicable to the facts of the decision. The effects of and problems with the WTO Panel's reasoning are discussed in a later section of the chapter. Our next task here is to explore the contours of the WTO regime.

5.2 ROOM FOR ENVIRONMENTAL AND HEALTH SAFETY PROVISIONS IN THE WTO

The WTO regime purports to facilitate international trade without discrimination.[522] Though the stated objective of the WTO complex is limited to the regulation of international movement of goods and services, it is often characterized as the engine of global administrative law.[523] The increasing reach of trade concerns in other international law realms at one level, and the difficulty in separating trade aspects from other legitimate issues, have effectuated an extremely strong presence of the trade regime in regulatory concerns of health, environment, ethics, human rights and development.[524] Obligations of Members with an obvious relevance to the regulation of GMOs primarily appear in GATT, 1994,[525] and the two WTO side agreements of TBT[526] and SPS.[527] All three Agreements, in principle, recognize legitimate exceptions towards a

---

[522] Core principles that were already enunciated in the original GATT of 1947 include the Most-Favoured-Nation principle (that appears in Article I), requiring Members to treat products emanating from all other Members in a similar manner; a general prohibition on quantitative restrictions, viz., quotas, embargoes and licensing schemes related to international trade of goods (Article XI); as well as the principle of national treatment (Article III). The latter requires Members to treat domestic products and imported products that are similar to the imported products in the same way. GATT 1994 incorporates GATT 1947 by reference, 31 Annex 1 A (1994) 33 *International Legal Materials* 1153.

[523] Sabino Cassese, 'Global standards for national administrative procedure' (2006) 68 *Law and Contemporary Problems* 109, 113–115.

[524] See Andrew T. Guzman, 'Global governance and the WTO' (2004) 45 *Harvard International Law Journal* 303.

[525] General Agreement on Tariffs and Trade, 1994, 1867 UNTS 187 (1994) 33 *International Legal Materials* 1125.

[526] The Agreement on Technical Barriers to Trade, 1994 (1994) 33 *International Legal Materials* 1140.

[527] The Agreement on the Application of Sanitary and Phytosanitary Measures, April 15, 1994, Annex 1A-4 to Final Act Embodying the results of the

variegated obligation for free movement of goods, in a conceptual recognition of national regulatory sovereignty. The nature and application of these exceptions substantially differ between these Agreements, though, in effect, these differences are papered over in the trade of GMOs, by the dominance of SPS over others through judicial interpretation in the *EC-Biotech* decision (examined in the next section).

The Dispute Settlement Body of the WTO, set up by the Dispute Settlement Understanding,[528] is the judicial arbiter of the fine line between, on the one hand, sovereign rights of Members to take appropriate measures to protect public morals, environment and health issues in their respective jurisdictions, and, on the other hand, arbitrary or discriminatory measures that restrict global trade. The DSB is widely recognized to have played a powerful and effective role in strengthening the trade regime into a ubiquitous and powerful instrument of global governance.[529] Though traditionally a focus on international trade negotiations and dispute resolution is the main avenue for understanding the global trade regime as an instrument of global governance, recent work on the functioning of expert committees within the WTO regime has revealed another important site where global rule-making affects Member regulation.[530] The work of the Codex Alimentarius, TBT and SPS Committees in the relevant global rule-making for GMOs is examined in a subsequent section to understand their effects on public participation in GMO regulation. The current section identifies the textual and interpreted principles in the aforesaid Agreements that are relevant to the protection

---

Uruguay Round of the Multi-lateral trade negotiations, (1994) 33 *International Legal Materials* 1153.

[528] Articles 1.1, 23 and Appendix 1 of Dispute Settlement Understanding on Rules and Procedure Governing the Settlement of Disputes, Annex 2 of the WTO Agreement (1994) 33 *International Legal Materials* 1226; Article 4.3 of the WTO Agreement and Article 11.1 of the SPS Agreement, and Article 14.1 of the TBT Agreement. See generally: Peter van den Bosche, 'World Trade Organization dispute settlement in 1997 (Part I)' (1998) 1 *Journal of International Economic Law* 161; Peter van den Bosche, 'World Trade Organization dispute settlement in 1997 (Part II)' (1998) 1 *Journal of International Economic Law* 479; John H. Jackson, 'Dispute settlement and the WTO: Emerging problems' (1998) 1 *Journal of International Economic Law* 329.

[529] See Jide O. Nzelibe, 'Interest groups, power politics and the risks of WTO mission creep' (2004) 28 *Harvard Journal of Law and Public Policy* 89, 90.

[530] Andrew Lang and Joanne Scott, 'The hidden world of WTO governance' (2009) 20 *European Journal of International Law* 575.

of environmental and health concerns even while principles of free trade are effectuated.

Members can pursue a range of policy concerns, enumerated in the general exceptions of GATT, 1994 as long as they are not applied in an arbitrary or unjustifiably discriminatory manner.[531] These exceptions include measures 'necessary to protect human, animal or plant life or health',[532] 'relating to the conservation of exhaustible natural resources'[533] and 'necessary to protect public morals'.[534] These exceptions are applied in conjunction with a number of related obligations in the *chapeau* of the Article, to ensure the use of such exceptions does not amount to arbitrary or unjustifiable discrimination between like goods or circumstances, and is not a disguised restriction on international trade.[535]

The general exceptions are examined in a WTO tribunal only if the respondent invokes them as a defence against a complaint, and subsequent to a demonstration by the complainant that there is a prima facie violation of a trade obligation. The respondent, in such situations, can declare that this supposed violation is in pursuance of a defined policy purpose falling within a specific exception.[536] It is for the defendant to establish that the impugned measure falls within the purview of the identified exemption. A plethora of lines of enquiries are used by WTO tribunals to ascertain the legality of these measures, including how two sets of products concerned are 'like goods', whether the imported goods are treated less favourably than these 'like goods', and whether such a treatment is necessary and permitted under any provided exceptions.[537] A two-stage enquiry is suggested by the Appellate Body in *US-Shrimp/*

---

[531]   A combination of Articles I, III and XI GATT seeks to unfetter restrictions on free trade and outlaw discrimination of like goods irrespective of national origins. This chapter does not attempt a detailed description of the central tenets of the WTO and GATT. Ample literature already deals with these issues that constitute an entire field of legal studies as International Trade Law. See Grant Isaac and William Kerr, 'Genetically modified organisms and trade rules: Identifying important challenges for the WTO' (2003) 26 *World Economy* 28, 29–32.

[532]   Article XX (b) GATT.

[533]   Article XX (g) ibid.

[534]   Article XX (a) ibid.

[535]   Article XX ibid.

[536]   Ibid. See further, Frieder Roessler, 'Appellate Body ruling in China-Publications and Audiovisual Products' (2011) 10 *World Trade Review* 119, 121.

[537]   Benn McGrady, 'Neccessity exceptions in WTO law: Retreaded tyers, regulatory purpose and cumulative regulatory measures' (2009) 12 *Journal of International Economic Law* 153; Christopher Doyle, 'Gimme shelter: The

Turtle – first, a substantive examination to determine whether the impugned measure falls within the specific exception.[538] Secondly, this is followed by a procedural examination of the manner in which the impugned measure is implemented by the Member to ascertain the necessity of the measure, and whether there are other measures reasonably available that entail lesser degrees of inconsistency with other provisions in GATT.[539]

Though the textual difference between 'necessary to' in health measures (in XX b) and 'related to' in conservation efforts (in XX g) is generally important, it becomes marginal in the case of international trade rules on GMOs because of the effect of the SPS Agreement on both these exceptions. This is because of, first, the principle that GATT must give way to specialised Agreements like SPS or TBT.[540] Secondly, the dominance of the provisions of the SPS Agreement is also effectuated by the decision of the Panel in *EC-Biotech*, an aspect that is discussed in detail in a subsequent section.[541] Related aspects of judicial interpretations of general exceptions are elaborated in greater detail in a subsequent section regarding the relevant exceptions on protection of public morality.[542]

---

necessary element of GATT Article XX in the context of the *China Audio-visual products* case' (2011) 29 *Boston University International Law Journal* 143.

[538] WTO Appellate Body Report, *US-Import prohibition of certain shrimp and shrimp products* WT/DS58/AB/R, adopted 12 October 1998. The complaint pertained to certain US restrictions on the import of shrimp that were not fished using a turtle excluder device. The Appellate Body recognized that the 'natural resources' in the exception of 'conservation of exhaustible natural resources' in XX (g) GATT include endangered species of animals or plants.

[539] Ibid.

[540] *General Interpretative note to Annex 1 A* of the WTO Agreement: 'In the event of conflict between a provision of the GATT and a provision of another agreement in Annex 1 A ... the provision of the other agreement shall prevail to the extent of the conflict'.

[541] There is a contrary possibility arising from the recent decision of the Appellate Body in *China-Audiovisuals* (n. 658), examined in the last section of this chapter. Here, the Appellate Body recognized the possibility of the use of Article XX defence in trade agreements other than GATT, thus making these defences possible even for a breach of SPS provisions.

[542] Text to nn. 661–667. It is instructive to recount an important remark by some prominent commentators, describing the pattern of judicial decisions on these exceptions, as one of strict interpretation in a way that trade rules trump domestic social objectives. See for instance, Steve Charnovitz, 'An analysis of Pascal Lamy's proposal on collective preferences' (2005) 8 *Journal of International Economic Law* 449, 467–469; Sanford E. Gaines, 'The WTO's reading

The two side agreements of SPS and TBT arose out of the concern that measures that prima facie appear to be for protection of the environment or health may in fact be disguised protectionism.[543] The relative achievement of minimizing obvious discriminatory protectionism led to a focus on Members' regulations that could potentially disguise protectionism in the name of safety. Such a situation necessitated additional legal instruments to ensure that the DSB does not get inundated with delicate and complicated balancing of Members' regulatory measures that apparently seek to protect human health, safety or the environment.

The SPS regime sets forth rights and obligations of Members with respect to any measure taken by a Member to protect the health or life of humans, animals or plants that may, directly or indirectly, affect international trade.[544] The letter and scheme of the SPS Agreement have science – scientific principles, process and evidence – as the yardstick to ascertain the legality of national regulatory measures.[545] There is a textual recognition of the Members' right to set their own appropriate level of protection, and the regulatory measures are sought to be disciplined by requiring them to be based upon a risk assessment, with a demonstrable and rational relationship between the risk assessment and the measure in question.[546] Further, such Member autonomy in setting the appropriate level of protection must be consistent across similar products, and cannot be either arbitrary or unjustifiably chosen for

---

of the GATT Article XX chapeau: A disguised restriction on environmental measures' (2001) 22 *University of Pennsylvania Journal of International Economic Law* 739.

[543]   Article XX GATT. The chapeau requires that the employment of general exceptions is subject to them being not a disguised restriction on international trade. See further, Gabrielle Marceau and Joel Trachtman, 'The Technical Barriers to Trade Agreement, the Sanitary and Phytosanitary Measures Agreement, and the General Agreement on Tariffs and Trade' (2002) 36 *Journal of World Trade* 811.

[544]   SPS Agreement, Preamble, Article 1.1 and Annex A.1. Pertinently the definition includes prevention or restriction of 'other damage' from the entry, establishment or spread of pests; Article 1.1 (d).

[545]   The SPS Agreement requires that any sanitary measure is based on scientific 'principles', is not maintained 'without sufficient scientific evidence' and that a risk assessment procedure is undertaken that takes into account the existing available information on toxicity and exposure. See Articles 2.2, 3 and 5.1.

[546]   Article 5.5 SPS.

particular products that may lead to an unjustified restriction on international trade.[547]

Importantly, the SPS Agreement provides for maintenance of provisional measures, if insufficiency of scientific evidence can be demonstrated.[548] These provisional measures are required to be based on available pertinent information, including harmonized global standards in SPS Committees and relevant international organizations, as well as SPS measures of other Members.[549] However, Members in the wake of such provisional measures are enjoined to seek additional information necessary for a new risk assessment to review its provisional measures within a reasonable period of time. It is important that the treatment of scientific uncertainty in this provision is generally considered weaker than versions of the precautionary principle found in international environmental law.[550] The Appellate Body in *Japan-Apples* emphasized that this provision is primarily intended to deal with new risks where little or no reliable evidence is available on the subject matter.[551] It underlined that the insufficiency condition is only fulfilled if 'the body of available scientific evidence does not allow, in quantitative or qualitative terms, the performance of an adequate assessment of risks as required'.[552] Such maintenance of provisional restrictions based on insufficiency of scientific evidence is examined in more detail in the next section.

The TBT Agreement is similar in its scheme to achieve a balance between regulatory measures and 'protectionist' discrimination in the realm of technical regulations; albeit with far less emphasis on scientific justification and identification of scientific agencies that shall produce standards for harmonization.[553] The balancing between national regulatory autonomy and checks against protectionism is configured in a different manner in the TBT Agreement, whereby technical regulations that are restrictive of trade are permitted as long as it is necessary to fulfil a legitimate objective such as protection of human health or safety,

---

[547] Ibid. See for more, Jeffery Atik, 'The weakest link: Demonstrating the inconsistency of "appropriate levels of protection" in *Australia-Salmon*' (2004) 24 *Risk Analysis* 483.

[548] Article 5.7 SPS.

[549] Ibid.

[550] Applegate (n. 366) 51–55.

[551] WTO Appellate Body Report, *Japan – Measures Affecting the Importation of Apples*, WT/DS245/AB/R, AB-2003-4 (2003) adopted 10 December 2003.

[552] Ibid para. 179.

[553] See Andrew Thomison, 'A new and controversial mandate for the SPS Agreement: The WTO Panel's interim report in the EC-Biotech dispute' (2007) 32 *Columbia Journal of Environmental Law* 287, 291–292.

animal and plant life, or health or protection of the environment.[554] It contrasts itself with the SPS framework through the absence of a formal and strict requirement of risk assessment and a scientific basis.[555] There is a process of international harmonization of standards envisaged through TBT Committees, and in cases where measures fall within both TBT and SPS Agreements the former is expected to defer to the latter.[556]

The SPS Agreement also embodies an ideal to promote global harmonization of these regulatory measures, through establishment of scientific standards, guidelines and recommendations by certain identified global agencies like Codex Alimentarius.[557] Here the default expectation is that national measures will adhere to the global standards, and any deviation ought to be consistent with the rigours of the scientific principle through a risk assessment. SPS discipline is supposed to have heralded a shift to a post-discrimination paradigm in international trade law, since it is applicable to all international trade, irrespective of GATT's injunction against discrimination. In other words, unlike GATT, where a violation of non-discrimination has to be established (for instance in Article XX b), SPS requirements of scientific rationality are necessary even without an a priori state of discrimination in trade.[558] Thus it is arguable that the ambit and impact of these side agreements (especially SPS) have gone far beyond the concerns about protectionism, towards an aspiration to universality, uniformity and singular reason to direct national regulations. Such disciplines, considered unique even within the trade regime, are

---

[554]   Article 2.2 of the TBT Agreement.

[555]   Prior to the Uruguay Round, the TBT Agreement, 1979 was the relevant legislation. It had limited effects: wherein, as long as a 'health regulation was necessary and was not a means of arbitrary or unjustifiable discrimination or poses a disguised restriction on international trade', such a measure was deemed legitimate; the Technical Barriers to Trade Agreement, 1979 (1979) 18 *International Legal Materials* 1079.

[556]   Article 1.5 TBT. See Lang and Scott (n. 530) for more on this aspect of TBT Committees.

[557]   See the Preamble, Article 3 and Annex A3 of SPS. See for an elaboration of this point: Vern Walker, 'Keeping the WTO from becoming the World Trans-science Organization: Scientific uncertainty, science policy, and fact-finding in the Growth Hormones dispute' (1998) 31 *Cornell International Law Journal* 251, 272–277.

[558]   See: Robert Hudec, 'Science and post-discriminatory WTO law' (2003) 26 *Boston College of International and Comparative Law Review* 185. See for an argument against this conventional position, Bernard Hoekman and Joel Trachtman, 'Continued suspense: *EC-Hormones* and WTO disciplines on discrimination and domestic regulation' (2010) 9 *World Trade Review* 151, 162–164.

criticized for having the effect of double-guessing and unduly restraining legitimate democratic choices, right at the outset of the regulatory process.[559]

Considerably earlier than the publishing of the Panel report in *EC-Biotech*, speculative anticipation of the decision brought much normative attention for the appropriate relationships between the SPS and other trade regimes *inter se*, as well as between trade regimes and protection of the environment and health.[560] Various commentators have correctly pointed out that institutional and other kinds of rigours vis-à-vis environmental and health regimes the world over are being tightened due to the widening ambit of the SPS Agreement.[561] Despite a number of earlier decisions, the expansion of SPS architecture into national regulatory geographies, over and above the other trade principles, clearly manifests itself in *EC-Biotech*. The Panel interpreted SPS measures broadly, and held many concerns that would traditionally be considered as an environmental issue as SPS measures. These include risks associated with consumption of GM pollen or plants by insects or wild animals (as a food risk to animal health), and damage to geochemical cycles or internal dynamics of populations of species (as 'other damage' from risks associated with pests).[562] Such expansion has been observed to make consideration of important factors like public opinion difficult, and also

---

[559]   Tracey Epps and Andrew James Green, 'The WTO, science and the environment: Moving towards consistency' (2007) 10 *Journal of International Economic Law* 285.

[560]   See for instance, Scott, 'European regulation of GMOs and WTO' (n. 209); Winham (n. 517); Isaac and Kerr (n. 531); Mystery Bridgers, 'Genetically modified organisms and the precautionary principle: How the GMO dispute before the World Trade Organization could decide the fate of international GMO regulation' (2004) 22 *Temple Environmental Law and Technology Journal* 171; Gavin Goh, 'GM food labelling and the WTO Agreements' (2004) 13 *Review of Community and International Environmental Law* 306; Robert Neff, 'The Cartagena Protocol and the WTO: Will the EU Biotech products case leave room for the Protocol' (2005) 16 *Fordham Environmental Law Review* 261.

[561]   See for instance, Christian Joerges and Julius Neyer, 'Politics, risk management, World Trade Organization governance and the limits of legalisation' (2003) 30 *Science and Public Policy* 219; Joanne Scott, 'On kith and kine (and crustaceans): Trade and environment in the EU and WTO' in Joseph Weiler (ed.), *The EU, the WTO, and the NAFTA: Towards a Common Law of International Trade?* (Oxford University Press 2000) 125.

[562]   *European Communities – Measures Affecting the Approval and Marketing of Biotech Products*, WT/DS291/R, WT/DS292/R, WT/DS293/R, Corr.1 and Add.1, 2, 3, 4, 5, 6, 7, 8 and 9, adopted 21 November 2006, paras 7.207, 7.292. See further, Lee, 'EU Regulation' (n. 15) 234–235.

to chill public pressure to actively pursue alternative trajectories to controversial technologies; including limiting the effect public pressure can have on the framing of environmental risk regulations.[563]

Though the WTO Panel is not expected to engage in a de-novo assessment of the matter, exigencies of judicial review usually go beyond the matter of veracity of formal procedures and the evaluation of the existence of manifest error. The Panel often examines questions like whether the studies or reports cited by a Member amount to a 'risk assessment' for the purposes of the SPS Agreement, whether 'available scientific evidence' is taken into account during the course of the assessment, and, even further, as to whether the scientific evidence that supports the impugned measures is sufficient.[564] In all the six cases in which SPS disciplines were discussed till date, WTO tribunals have employed a four-pronged enquiry:[565]

---

[563]   See for an apt elaboration of this point Jacqueline Peel, 'A GMO by any other name … might be an SPS risk! Implications of expanding the scope of the WTO Sanitary and Phytosanitary Measures Agreement' (2006) 17 *European Journal of International Law* 1009. Also see, Thomison (n. 553) 307: 'with its demanding requirements for SPS measures and now a broadened mandate, the SPS agreement appears increasingly likely to surface on the WTO scene as the premier instrument for attacking disputed health and environmental protection measures in the global trade arena'.

[564]   See *EC-Hormones I* (n. 426) para. 589. See for an examination of whether the requirement of risk assessment is procedural or substantive under the SPS process in this decision: Jacqueline Peel, 'Risk regulation under the WTO SPS Agreement: Science as an international normative yardstick?' (Jean Monnet Working Paper 02/04) http://www.jeanmonnetprogram.org/papers/04 last accessed 19 October 2009.

[565]   These are: *EC-Hormones I* (ibid.); WTO Appellate Body Report *Australia – Measures Affecting Importation of Salmon*, WT/ DS18/ AB/ R, AB-1998-5 (1998) adopted 20 October 1998; WTO Appellate Body Report *Japan – Measures Affecting Agricultural Products*, WT/DS76/AB/R, AB-1998-8 (1999) adopted 22 February 1999; *Japan-Apples* (n. 551); *EC-Biotech* (n. 562); *Canada/United States – Continued Suspension of Obligations in the EC – Hormones Dispute*, WT/DS320/AB/R, WT/DS321/AB/R, adopted 14 November 2008 (*EC-Hormones II*). All the decisions appear to have involved various degrees of substantial examination of the regulatory measures. See further: Joost Pauwelyn, 'The WTO Agreement on Sanitary and Phytosanitary (SPS) measures as applied in the first three SPS disputes EC-Hormones, Australia-Salmon and Japan Apples' (1999) 2 *Journal of International Economic Law* 641, 644.

1. identifying the scientific basis upon which the measure was adopted (through which the issue of whether the measure in question is an SPS measure is also investigated);
2. examining the veracity and credibility of the source of the study, including an examination into whether international guidelines (like CODEX guidelines) and other Members' risk assessments have been taken into account;
3. an assessment of the coherence and objectivity of the cited scientific study, so as to accept it as a risk assessment; and
4. an assessment of whether there is a rational relationship between the findings of the scientific study and the SPS measure in question.[566]

The next section examines the relevant findings of the Panel in *EC-Biotech* to investigate the room for public contestations and participation in regulation of technology in WTO law. This is pertinent because it involved important interpretations regarding the meaning and ambit of 'scientific uncertainty' in the regulation of GMOs, and the application of the precautionary principle in it. The milieu of dispute, including the facts, circumstances and parties in the case, makes its examination directly relevant for our purpose.

5.3 EXAMINING THE EC-BIOTECH DECISION: EMPHASIS ON 'SOUND SCIENCE'?

The dispute in *EC-Biotech* was initiated after a series of complaints about the obstacles to the approval of a number of GM agricultural products in the EU, which were sought to be imported from the USA, Canada and Argentina.[567] Howse aptly encapsulates the broad context of the *EC-Biotech* dispute at the WTO:

> Not only did the Panel have to take a stand on the limits of science, or technocratic regulatory controls to protect against objective risk, but in this regard faced more complex issues than ever addressed before by an adjudicating body. The dispute also concerned an extremely charged political issue, partly because of inherent ethical sensitivities with regard to foodstuffs, partly due to public skepticism about the role of science, and partly due to a

---

[566] See further, Catharina Koops, 'Suspensions: To be continued, the consequences of AB Report in *Hormones II*' (2009) 36 *Legal Issues of Economic Integration* 353, 361.
[567] *EC-Biotech* (n. 562).

common public perception of the complaint as being driven by the interests of an untrustworthy industry.[568]

The Panel took almost thirty months to issue its final report, as opposed to the usual provision for six months (or, at the very most, nine months in exceptional circumstances),[569] giving a glimpse of the intensity and complexities of the GM debate in agriculture.[570]

The main issues identified by the panel for adjudication were limited to procedural questions. These were:

1.   whether there was a general moratorium on GM products between October 1998 and August 2003 in the EC (as it then was called);
2.   alternatively, whether there was a product-specific moratorium in the same period;
3.   whether either of these EC measures was illegal; and
4.   whether the twenty-seven impugned Member State safeguard measures against specific products earlier approved by the Community regime violated the WTO regime.

The last issue was accompanied by a technical question as to whether specific scientific studies that were cited to justify the safeguard measures could be accepted as risk assessments within the ambit of the SPS Agreement. Having found the existence of a general moratorium on GM products, the Panel held all the aforementioned measures to violate various provisions of the SPS Agreement. In the face of such violations, the Panel found it unnecessary to examine violations regarding the other agreements of the WTO regime arising from the same measures. Given the fact that the dispute was about procedural issues, and that too pertaining to the earlier regulatory regime, commentators have pointed out that the decision invalidating EU and Member State measures has not

---

[568]   Robert L. Howse and Henrik Horn, 'European Communities – Measures affecting the approval and marketing of biotech products' (2009) 8 *World Trade Review* 49.

[569]   Article 16 Dispute Settlement Understanding (n. 528).

[570]   The complainants are among the four biggest growers of biotech crops. In 2005, the US accounted for 55 per cent of the global area planted with biotech crops (49.8 million hectares), Argentina 19 per cent (17.2 million hectares) and Canada 6 per cent (6.1 million hectares); cited in Clive James, 'Global status of commercialized biotech/GM crops: 2006' (International Service for the Acquisition of Agri-biotech Acquisitions, ISAAA Brief # 35, 2006) <http://www.isaaa. org/resources/publications/briefs/35/pptslides/Brief35slides.pdf> last accessed 26 May 2011.

particularly altered the current EU framework (despite initial alarmist commentary).[571]

Cheyne identified a number of important issues left unresolved by the Biotech Panel, including the appropriate role of the precautionary principle in WTO law, and also the relevance of the doctrine of substantial equivalence in the trade regime (in other words, whether GMOs and conventional products can be considered like products for the purposes of GATT).[572] There are a number of pertinent issues in the decision that are relevant for an examination of the room WTO rules allow for regulatory measures which seek to improve representational and participatory issues during regulation of GMOs. First among these is the Panel's definition of a risk assessment, and how such a definition affects the understanding of scientific uncertainty and sufficiency, elaborated in the next subsection.

5.3.1 What is a Risk Assessment?

The Panel in *EC-Biotech* sought to employ a set of criteria previously set by the Appellate Body in *Australia-Salmon*[573] to identify the necessary elements of what could be considered a risk assessment.[574] This identification was necessary for the argument that the impugned measures fall within the ambit of SPS measures.[575] By setting a high definitional

---

[571]   See, for a useful treatment of this point, as also for a very good commentary on the Panel's decision: Denise Prevost, 'Opening Pandora's Box: The Panel's findings in the EC-Biotech Products dispute' (2007) 34 *Legal Issues of Economic Integration* 67. The complainants in any case shied away from substantially challenging the EC GMO regulatory regime per se, but concentrated on the delays that the regime generated. Canada even argued that the EC would not have erred in any manner if it had adhered to its own assessment and approval procedure.

[572]   Under Article III GATT and Article 2.1 TBT. Ilona Cheyne, 'Life after the Biotech Products dispute' (2008) 10 *Environmental Law Review* 52, 64.

[573]   The three elements considered essential by the Appellate Body in *Australia-Salmon* were: '1) *identify* the diseases [or pests] whose entry, establishment or spread a Member wants to prevent on its territory, as well as the potential biological and economic consequences associated with the entry, establishment or spread of these diseases; 2) evaluate the *likelihood* of entry, establishment or spread of these diseases [or pests], as well as the associated potential biological and economic consequences; and 3) evaluate the likelihood of entry, establishment or spread of these diseases [or pests] *according to the SPS measures which might be applied*', *Australia-Salmon* (n. 565) para. 121.

[574]   A combination of Article 2.2 and Annex A (4) of the SPS Agreement expects these measures to be based on risk assessments, text to nn. 544–550.

[575]   *EC-Biotech* (n. 562) para. 7.3040.

threshold to employ this definition, the Panel refused to hold various scientific studies as risk assessments. These cited studies could have cast important doubts about the sufficiency of scientific knowledge for regulation, and could even negate the assumption of the previous risk assessments.

For instance, the Panel rejected several studies on lacewing larvae that confirmed the harmfulness of *Bacillus thuringiensis* (Bt.) toxin residues in GM maize for lacewings (predatory insects that do not feed on maize), as they were undertaken under laboratory conditions. The logic of the Panel was that in non-laboratory conditions the lacewings would have had a choice between Bt. corn-fed prey, and prey not raised on genetically modified corn, and if larvae chose the latter, the Bt. toxin residue would not have shown in the studies.[576] This reasoning is an extension of the differentiation between laboratory conditions and 'real world' conditions earlier made by the Appellate Body in *EC-Hormones I*. The Appellate Body in this case held that 'not only risk ascertainable in a science laboratory operating under strictly controlled conditions, but also risk in human societies as they usually exist' must be taken into account.[577] The use of this distinction between laboratory conditions and 'real world' conditions by the Panel in *EC-Biotech* is contrary to the legal form and intent of its original employment in *EC-Hormones I*. The latter recognized the possibility of risks existing in non-laboratory conditions that might not be visible in laboratory conditions. However, such an observation cannot be used to negate the consideration of hazards that were identified in laboratory studies, by a counter-claim that hazards identified in laboratory conditions cannot be accepted, as they were not tested in 'real world' conditions.[578]

---

[576]   Ibid, para. 7.3098.

[577]   *EC-Hormones I* (n. 426) para. 187.

[578]   However, it is pertinent that the Appellate Body in its final analysis still insisted on a high degree of specificity in the scientific studies examining consumption of meat from cattle treated with artificial hormones, seeking to establish a general risk of cancer from such consumption, *EC-Hormones I* (n. 426) para. 198. Lee cited Sykes' reminder of this insistence by the Appellate Body that places an insurmountable hurdle to regulate risks of this nature, one which cannot be demonstrable through particularized scientific studies. This effectively stipulates which risks the Members must tolerate and which they may select to avoid. Lee opined that notwithstanding the lack of clarity, it is implausible that the Appellate Body wishes to deny even the possibility of such Member regulation, (n. 15) 216. Notwithstanding the plausible intentions of the Appellate Body, Sykes' characterization of effects of such insistence being not much more than window dressing appears apt, given difficulties of inherent

Herwig characterized the rejection of a lacewing study by the *EC-Biotech* Panel as counterfactual, since the study could be easily accepted as a risk assessment because it is a direct, dose-dependent risk assessment for the specific risk that is likely to be encountered under 'real world' circumstances.[579] The restrictive manner in which the Panel had gone about rejecting laboratory-based scientific studies as not being a risk assessment is criticized as unrealistic and unnecessarily harsh:

> A study that called for further evaluation did not constitute a risk assessment, and neither did a study which was designed to help set targets for future studies. A study which made reference to the possibility of a risk but made no evaluation of it likewise did not constitute a risk assessment. A laboratory based study which suggested possible effect was not sufficient; nor was a study which claimed to evaluate the likelihood of spread without differentiating between GM and conventional crops. A document which referred to studies and their conclusions but did not actually contain those studies did not constitute a risk assessment. Nor was a study which dealt with Bt. proteins but was not specific to Bt. 176 maize sufficient ... *However in situations of real uncertainty and where scientific studies raise doubts about the validity of assumptions made in previous studies, requiring detailed evaluation of likely risks may be overly strict.*[580]

This negation of specific scientific studies was crucial to the whole scheme of the Panel's legal reasoning. Not only did such a rejection remove the scientific basis from the impugned measures, a central requirement of legality under the SPS Agreement,[581] but it also negated an assertion that these studies point towards a situation of insufficiency of scientific evidence.

5.3.2 Treatment of 'Insufficiency' of Scientific Evidence

Importantly, the legality of the impugned EU Member State safeguard measures hinged upon an evaluation of the sufficiency of scientific evidence that is closely connected to the judicial determination of the existence of prior risk assessments. The Panel held that the mere existence of previous EFSA assessments, and subsequent reassessments

---

scientific incertitudes in such situations. See further: Alan Sykes, 'Domestic regulation, sovereignty and scientific evidence requirements: A pessimistic view' (2004) 3 *Chicago Journal of International Law* 353, 364.

[579] Alexia Herwig, 'Whither science in WTO dispute settlement' (2008) 21 *Leiden Journal of International Law* 823, 837.

[580] Cheyne (n. 572) 57. Emphasis added.

[581] SPS Agreement (n. 527).

by Member States, demonstrated that scientific evidence was no longer insufficient, thus impugning the legality of the safeguard measures under Article 5 (7) SPS.[582] Here, the Panel employed the logic that an assertion of insufficiency in scientific evidence to make a risk assessment cannot be invoked in situations where risk assessments are already carried out.[583] Such a position is quite oblivious to the fact that insufficient evidence is a subjective concept, relative to the Members' level of protection. New scientific evidence or information could sufficiently question central assumptions of prior risk assessments, but nevertheless be insufficient to carry out a new risk assessment, based on the Members' different level of protection. Further, it is not realistic to expect that an existing risk assessment, even for the same level of protection, can only be trumped by definite, falsifying evidence and not by preliminary evidence of lesser corroboration.[584]

Such an interpretation of insufficiency borders on incomprehension of the inherent ambiguities with which scientific enquiries of this nature have to work, particularly during the development of radical technologies. Related to the discussions about the treatment of scientific uncertainty within expert systems in Chapter 4 is the Panel's approach of categorizing situations of sufficiency and insufficiency as mutually exclusive, as soon as a risk assessment exists.

> The distinction between the two hinges primarily on whether data allows for applying risk assessment methods to determine the risk and not on whether there are scientific doubts regarding the validity of the risk assessment method or its underlying theories or on whether preliminary scientific evidence suggests a potential hazard. Once a framing of risk is used, in other words, the rationality of decisions about hazards is determined by that frame and no longer by any persisting uncertainties.[585]

---

[582]   *EC-Biotech* (n. 562) paras 7.3260, 7.3273, 7.3286, 7.3300, 7.3314, 7.3327, 7.3341, 7.3356, 7.3369.

[583]   Ibid.

[584]   This position, deducible from a reading of the decision, is further elaborated by the Panel in its subsequent letter to the parties explaining the concept of insufficiency: 'scientific evidence which is at one point sufficient for a risk assessment can later become insufficient if new scientific evidence becomes available that negates the validity of the existing risk assessment but is itself insufficient to perform a risk assessment'. *EC-Biotech* (n. 562) Annex K.

[585]   Herwig (n. 579) 834.

Here the Panel emphasized that a Member can take into account factors that affect scientists' level of confidence in a risk assessment they have already carried out, only in determining the measures to be applied for achieving its appropriate level of protection.[586] Herwig correctly identifies the implication of this:

> [O]nce scientists consider evidence before them to be a risk assessment, they may absorb uncertainty into risk through levels of confidence. But would they also accept that there can be uncertainty which cannot be absorbed into a framing of risk? Hardly, because this would call into question their conclusion that the evidence before them is a risk assessment – that is, an assessment that really does tell us something about the probability of a hazard.[587]

She concluded that the dialectical nature of risk as interwoven with uncertainty was neglected, thereby over-emphasizing the role of science in risk regulation at the expense of normative or ethical elements.[588]

Further, a Member's right to set its own appropriate level of protection (recognized in Article 5.5 SPS) comes to the fore. The need to recognize that different levels of protection may correspond to different levels of determinacy of insufficiency is crucial here. This aspect of difference in level of protection affecting the determination of insufficiency was later recognized by the Appellate Body in *EC-Hormones II*.[589] Here it was held that scientific evidence which is used to conduct an existing risk assessment may be found as insufficient by another Member who sets a higher level of protection. This is since the higher level of protection may require analysis of different parameters of hazard, affecting the scope and method of the risk assessment.[590] However Members' right to fix higher

---

[586]   *EC-Biotech* (n. 562) paras 7.3234–7.3246.

[587]   Herwig (n. 579) 834.

[588]   Ibid. 'To the extent that uncertainty remains intertwined with risk and cannot always be resolved by scientific methods, decisions about appropriate regulation of risk must also involve normative or ethical judgment'. The Panel 'neglected this by holding that the protection purposes of the regulator were not relevant for determining the insufficiency of scientific evidence within the meaning of Article 5(7) of the SPS Agreement'.

[589]   *EC-Hormones II* (n. 565).

[590]   We avoid a detailed discussion regarding the *EC-Hormones II* judgment, though pertinent findings that are relevant for us are referred to at appropriate places. The dispute was brought to the DSB by the EC seeking suspension of countermeasures by the US and Canada, originally instituted in retaliation to the continuation of SPS measures that were invalidated in *EC-Hormones I*. In *EC-Hormones II*, the EC unsuccessfully argued before the Panel that the countermeasures were now unauthorized restrictive measures, citing seventeen

levels of protection is subject to further discipline in the trade regime; where, if in one product the level of acceptable risk is kept high, while in another product of similar situation, it is kept low, such a situation may be found to be disguised discrimination.[591]

The Panel in *EC-Biotech* identified four requirements that must be met for a Member to use provisional measures under Article 5.7 SPS. These are:[592]

1.   a demonstration that the relevant scientific information is insufficient,
2.   the provisional measure is based on pertinent available information,
3.   the Member should seek additional information necessary for a more objective risk assessment, and
4.   the Member should review the assessment within a *reasonable* period of time.

Thus the implication here is that a condition of insufficiency of scientific information is not enough for the employment and continuation of such provisional measures. Rather, according to the Panel's interpretation, subsequent research and review within a reasonable time-frame has a gate-keeping function for the invocation of these provisions. Some commentators have criticized this clubbing of the obligation for further

---

scientific studies commissioned by it as the basis of the current continuation of the ban against beef hormones. Upon cross-appeals, the Appellate Body found that the Panel failed to make an objective assessment of the matter before it, in its consultations with scientific experts, misinterpreted Articles 5.1 and 5.7 of SPS, misallocated the burden and applied an incorrect standard of review. Given such a Panel report, the Appellate Body found it difficult to complete the review, and returned the matter to the Panel for further consideration. See further, Koops (n. 566); Hoekman and Trachtman (n. 558).

[591]   This scenario is amply visible in the context of the *Australia-Salmon* decision (n. 565). Atik concludes that readings of different situations under Article 5.5 SPS, which permit different levels of protection for different products, are similar in the way interpretations of like products dominate the findings of discrimination under GATT Art. III; Atik (n. 547) 489. Here, however, the likeness of the situations or products only has to 'present some common element or elements sufficient to render them comparable', and is very broadly construed, as opposed to the restrictive nature of GATT, which requires *likeness or direct competitiveness or substitutability* between products to be established; *EC-Hormones II* (n. 565).

[592]   *EC-Biotech* (n. 562) paras 7.3234–7.3235. These criteria were originally set out by the Appellate Body in *Japan Varietalsby the WTO academics (Japan – Measures Affecting Agricultural Products)*, para. 179.

research to the right itself as erroneous.[593] They point out that tying the right to use measures under situations of insufficiency of scientific information to subsequent research and review is not textually permitted in the SPS Agreement. Thus they argue that the right to take provisional measures is recognized under the Agreement for all cases where relevant scientific evidence is insufficient, and the obligations for further research and review are supplementary obligations, whose breach invites the usual sanctioning mechanisms, without the denial of the exercise of the original right.[594] Further, as mentioned earlier, unlike the Appellate Body's rationale in *EC-Hormones II*, the Panel in *EC-Biotech* appears to have erred in denying the implicit connections between different levels of protection and insufficiency. The applicability of the precautionary principle in the SPS Agreement is related to this problematic treatment of insufficiency of scientific information, sometimes recognized as a watered-down version of the principle itself.[595]

5.3.3 Applicability of the Precautionary Principle

There were two possible legal routes of engagement with the principle available to the Panel in *EC-Biotech*, viz., first the principle as *jus cogens*, or an enquiry as to whether the principle is part of customary international law, and, secondly, the applicability of the principle as employed in the Cartagena Protocol. The Panel continued the trend among case law in the WTO framework of avoiding the question of the general applicability of the precautionary principle.[596] The Panel held that it was unnecessary to give a finding on the legal status of the principle as *jus cogens*, even while acknowledging that if the precautionary principle is part of customary international law it must be taken into account in the general interpretation of WTO law.[597]

---

[593] Andrew Lang, 'Provisional measures under Article 5.7 of the WTO's agreement on Sanitary and Phytosanitary measures: Some criticisms of the jurisprudence so far' (2008) 42 *Journal of World Trade* 1085.

[594] Ibid 1091–1095.

[595] Applegate (n. 366).

[596] *EC-Biotech* (n. 562) para 7.89. This was in line with the Appellate Body's finding in *EC-Hormones II* that the status of the precautionary principle is an ongoing debate, despite its incorporation in a number of international treaties, and it would be imprudent for the Appellate Body to take a position on the same; *EC-Hormones II* (n. 565) para. 121. See for the general trend: Prevost (n. 571).

[597] *Qua* Art.31.3 (c) of the Vienna Convention on the Law of Treaties, 1969, (1969) 8 *International Legal Materials* 679.

Further, the Panel avoided the question of employing the precautionary principle as expressed in the Cartagena Protocol by holding the Protocol itself as inapplicable to the legal context of the dispute, since a number of complainants had not ratified the same.[598] The legal issue identified as key was whether the reference to 'the parties' in Article 31(3)(c) of the Vienna Convention on the Law of Treaties (internationally recognized as applicable law) applied only to the disputing parties or alternatively to the whole membership of a multilateral treaty. If 'the parties' referred to here were merely the disputing WTO Members, a second treaty concluded between some of the disputing WTO Members only and not the whole WTO membership could be considered in the interpretation of the SPS Agreement. Alternatively, if 'the parties' referred to all Members of the WTO, only treaties with an identity of membership with the WTO agreements could be referred to as an outside interpretative source within the Convention.

The Panel held that only treaties concluded between all WTO Members were an outside interpretative source and therefore declined to consider the Biosafety Protocol. After taking notice of the Appellate Body ruling in *US-Shrimp* that 'WTO law cannot be read in clinical isolation from public international law', it held that such a drawing from other rules of international law was not an obligation but merely a discretionary aid to inform and interpret SPS provisions.[599] The reasoning used to exclude the Cartagena Protocol, including the specific expression of the precautionary principle in it – that such an agreement can only be considered as relevant if all WTO Members have signed and ratified the specific environmental agreement – poses grave concerns regarding the applicability of multilateral environmental agreements.[600]

---

[598]   *EC-Biotech* (n. 562) paras 7.71, 7.72.
[599]   *EC-Biotech* (ibid) para. 7.94.
[600]   See Prévost (n. 571) 86–93. Further, Article 24 of the Cartagena Protocol stipulates the trans-boundary movements of LMOs between parties and non-parties to be in consistent with the objective of the protocol, where parties may enter into bilateral, regional and multilateral agreements with non-parties for this end, and also exhorts parties to encourage non-parties to adhere to the substance of the Protocol.

This finding has been criticized also on textual grounds,[601] raising difficult issues of fragmentation of international law.[602] The UN International Law Commission expressedly criticized this stand of the Panel, and sought to address this important issue:

> Bearing in mind the unlikeliness of a precise congruence in the membership of most important multilateral conventions, it would become unlikely that any use of conventional international law could be made in the interpretation of such conventions. This would have the ironic effect that the more the membership of a multilateral treaty such as the WTO covered agreements expanded, the more those treaties would be cut off from the rest of international law. In practice, the result would be the isolation of multilateral agreements as 'islands' permitting no references *inter se* in their application.[603]

For the reasons discussed above, it would appear that the science-based requirements that the SPS framework textually demands are sought to be judicially extended not only to other trade realms like the TBT Agreement, but also to the environmental realm. Notwithstanding the caution the Panel appears to employ by stressing that it has not determined 'whether biotech products in general are safe or not', the framing of the ambit of the SPS framework in an expansive way is found to have far reaching and potentially debilitating effects on the regulation of environmental risks.[604]

---

[601]   Herwig (n. 579) 837, noted the relevance of Article 2(1) (g) of the Vienna Convention, which maintains that 'party' means a state which has consented to be bound by the Treaty and for which the Treaty is in force. Since this 'says nothing about disputing parties, some of or the whole membership of a multilateral treaty', she asserts that Article 31(3) (c) read in conjunction with this definition merely tells us that the international treaty binding even only the disputing parties, but not the whole WTO membership would comply with the provisions of the Vienna Convention.

[602]   See for instance: Joost Pauwelyn, *Conflict of Norms in Public International Law* (Cambridge University Press 2003) ch. 3–5; Matthew Craven, 'Unity, diversity and fragmentation of international law' (2003) 14 *Finnish Yearbook of International Law* 3.

[603]   International Law Commission, 58th Session, *Fragmentation of International Law: Difficulties Arising from the Diversification and Expansion of International Law*, Report of the Study Group of the International Law Commission, A/CN.4/L.682, 13 April 2006 para. 471.

[604]   Peel, 'A GMO by any other name' (n. 563), identifies the logic used by the Panel to expand the ambit of the SPS Agreement beyond the conventional understanding of the realm of the SPS agreements into the realm of not only the

As elaborated in the third chapter, once we recognize the political nature of risk management over and above the normativities employed in the stage of risk analysis, the multiple possibilities to characterize the contours of (in)sufficiencies of scientific evidence, and their connections to the levels of protection, come to the fore. Here, the importance of participation to choose between these multiple possibilities is affected by the Panel's interpretation of insufficiency. Further, many critics of the Panel decision in *EC-Biotech* highlight the concerns of fragmentation of international law. Whether such a concern of fragmentation can be solved entirely through expert consensus is an issue that will be addressed at the end of the next section.

5.4 SPS COMMITTEES, CODEX STANDARDS AND LIMITS OF COOPERATION

Traditional foci on DSB interpretations about the legitimate ambit of national regulatory sovereignty, and on high profile international trade negotiations, are said to hide considerable amounts of cooperative governance through the expert committees. Lang and Scott suggest that a keener focus on (the more than thirty-five) committees and the emergence of interpretative communities that serve to elaborate the (partly) open-ended norms laid down in the relevant agreements is necessary.[605] This, they contend, reveals a picture of the WTO that is more dynamic, cooperative and reflexive than conventional understandings of the WTO regime. They emphasize that such a focus on this world of apparently mundane activities in the expert committees reveals an administrative infrastructure which produces technocratic consensus towards supervision, monitoring and management of at least some important aspects of governance in the trading system.[606] In the context of this book, this section seeks to clarify the contours of possible EU deviation from

---

TBT Agreement, but also of multilateral environmental agreements. She convincingly argues that the textualist reading of the SPS provisions is not only patently incorrect, but also that it has undesirable implications for the development of national and international environmental regulatory regimes. For an argument regarding the lack of institutional capacity of WTO process to consider non-market values in general, and specifically how *EC-Biotech* fails to consider cultural factors, see: Layla Zurek, 'The European Communities Biotech dispute: How the WTO fails to consider cultural factors in the GM food debate' (2006) 42 *Texas International Law Journal* 345.

[605]   Lang and Scott (n. 530).
[606]   Ibid 576.

classical risk parameters, which facilitate public participation through the activities of these committees.

The SPS Committee is intended to provide a regular forum for Members to raise related 'specific trade concerns' and to monitor harmonization of international standards by relevant organizations like the Codex Alimentarius Commission (Codex).[607] Members are required to notify their SPS measures in the Committee so that others can raise concerns about them. This helps to detect disagreements at an early stage, and may resolve them through broader sharing of information, better understanding of each others' positions and building of research capacities. Such a state of better mutual understanding could lead to possible withdrawals or adjustments of contentious measures. Decisions in this regard can only be taken if no Member formally objects, and commentators feel that the Committee has generated greater empathy between Members and a heightened awareness of the needs and difficulties of other states.[608]

Generally, though committee procedures can possibly minimize conflict, the limits of such cooperation are stark in situations of incommensurable differences. In cases like GMOs or regarding the general ambit of the use of the precautionary principle, these committee spaces are seen to have become sites where existing sharp differences are rehearsed, where competition to gain regulatory allies among Members rather than to bridge positions is the order of the day.[609] Existing differences are seen to be rehearsed in some important areas during Codex proceedings (see the next subsection). This is visible in the experience of the TBT Committee as well, where EU insistence on labelling has been an issue of consistent controversy.[610] Steinberg reminds us of the acute possibility of committee members strategically providing incomplete or incorrect information, which makes existing imbalances in information between Members particularly stark.[611]

The SPS Committee is also a site where open-ended norms laid down in the Agreement are elaborated; thus the nature of membership becomes

---

[607] See Articles 12, 3.3 and 5.5 of SPS Agreement.

[608] Lang and Scott (n. 530) 594.

[609] Richard H. Steinberg, 'The hidden world of WTO governance: A reply to Andrew Lang and Joanne Scott' (2009) 20 *European Journal of International Law* 1063, 1064.

[610] Nico Krisch, 'Pluralism in post-national risk regulation: The dispute over GMOs and trade' (2010) 1 *Transnational Legal Theory* 1, 27. This Committee is established by Article 13 TBT.

[611] Steinberg (n. 609).

important. All WTO Members have automatic membership, usually represented by trade diplomats or particular ministry specialists. While various intergovernmental bodies are granted different types of observer status, importantly non-governmental observers are simply not admitted.[612] Lang and Scott suggest how these activities of the Committee contribute to the emergence of interpretative communities, and such trans-governmental networks need not necessarily amplify democratic norms, process or concerns. A further examination of the work of the Codex Alimentarius Commission throws better light on this aspect of the impediments to the expression of public contestations in and through the functioning of expert committees/epistemic communities.

5.4.1 Codex Alimentarius Commission

An important aspect of the health–environment–trade complex of global rules on GMOs is furthered through the functioning of the Codex Alimentarius Commission.[613] As a joint endeavour of the Food and Agriculture Organization (FAO) and the World Health Organization (WHO) earlier charged with the formulation of international food codes as a voluntary standard-setting process, Codex pre-dates the WTO Agreements by more than three decades.[614] However, the role and functions of the Commission have considerably transformed since the incorporation of its work within the ambit of the SPS, as one of the five

---

[612]   Standard-setting bodies like Codex and the World Organization for Animal Health, and a number of intergovernmental bodies like UNCTAD, the WB, IMF and FAO are granted permanent observer status, while other bodies like EFTA, the OECD and ACP have been granted ad-hoc status; Lang and Scott (n. 530) fns 70, 71.

[613]   Article 1 (a) Statutes of the Codex Alimentarius Commission, *Commission Procedural Manual 3* (14th edn, Codex Alimentarius Commission 2004) <ftp://ftp.fao.org/codex/ProcManuals/Manual14e.pdf> last accessed 21 November 2011. Article 1 states the purpose of the constitution of the Joint FAO/WHO Joint Food Standards Program.

[614]   The scientific assessment and consensus-building of this Commission consist of various horizontal practice-specific standards, including ones on hygiene and manufacturing, and limits on pesticide residues in food, as well as commodity-specific standards, like in packaging and processing. See for more on the Commission: WHO/FAO, *Understanding the Codex Alimentarius* (FAO 2005) <www.fao.org/docrep/008/y7867e/y7867e/y7867e00.htm> last accessed 21 November 2011.

international standards bodies whose standards could legitimate expectations of either compliance or exceptions to free trade rules under it.[615]

There is a twofold quasi-delegation of the mandate of the SPS Agreement, first to harmonize and develop uniform standards, and secondly to monitor and discuss the deviations from these standards by Members, in a kind of peer review process. Any deviation from the recommendations of the Codex requires a justification by Members through a risk assessment involving 'sufficient scientific evidence'.[616] The DSB has also used Codex standards in disputes relating to TBT, for instance the compatibility of an EC measure with Article 2.4 of TBT was evaluated by reference to a 1978 Codex standard in *EC-Sardines*.[617]

This reliance on an international body, which was until then relatively obscure, provided the WTO architecture with a universalizing normative authority of scientific expertise in its goal of rationalizing food safety regulation throughout the globe, notwithstanding possibilities of significantly divergent risk constructions among Members. The inclusion of Codex within SPS facilitated a scientific universalism through risk analysis for the WTO regime, since there is a default obligation that Members will rely on the standards set by Codex in their SPS measures. Though deviations from Codex standards are textually permitted as long as they are intended to achieve a higher level of protection, the Panel's decision in *EC-Biotech* reminds us of the considerable difficulties in employing them. Commentators have identified fundamental ways in which this recognition by the WTO regime has changed Codex, as if there is a new body that is doing its work.[618] Recent guidelines show an inclination to issue meta-guidelines, while prior to the mid-1990s its standards were relatively narrow in scope, dealing with one procedure or even one product at a time. Publication of a set of procedures and principles related to risk analysis of goods in general, and risk analysis of foods derived from modern biotechnology, would have been considered

---

[615] Article 3 read with Annex A. 3 (a) SPS. The other bodies include the World Organization for Animal Health, and the International Plant Protection Convention.

[616] Articles 3.2 and 5 of SPS.

[617] WTO Panel Report *European Communities – Trade Description of Sardines* WT/DS231/Panel/R adopted 29 May 2002 and WTO Appellate Body Report WT/DS231/AB/R adopted 26 September 2002.

[618] Sarah Poli, 'The European Community and the adoption of international food standards within the Codex Alimentarius Commission' (2004) 10 *European Law Journal* 613. She termed Codex as 'a satellite organization of the WTO', 615.

too far-fetched before the Authority was placed within the network of free trade policy actors.[619]

The standards formulated within Codex were entirely voluntary before the grant of coercive authority to these standards by giving incentives in the WTO trading regime for Members to base their regulations on them. The shift from a strictly voluntary organization to a global agency with quasi-legal authority is said to have transformed the body in important ways. As mentioned earlier, recognition of Codex within the WTO regime has coincided with the movement from a voluntary scientific standard-setting agency to a particular emphasis on probabilistic processes based on risk analysis. Codex itself recognized that this new status required several important changes in its own practices to foster a harmonized approach through making it 'consistent with science-based risk assessment'.[620] Winickoff and Bushey recall how employment of risk analysis in Codex, from a probabilistic analytical technique that was not very frequently used, transformed itself into the very grammar of decision-making of Codex.[621] The point they make is one of co-production of normativity, where the trading regime, by claiming to adopt pre-existing science-based standards at the international level, was in fact lending its legal and executive power in the WTO legal regime to transform Codex into a global agency, and necessitating it to generate such standards. The mandate for, and the self-imagination of, risk analysis, transformed the agency in its development of new norms and practices for management of knowledge, expertise and evidence in regulatory decision-making, and in turn gave the trading regime a claim to technical neutrality, universal normativity and legitimacy.[622]

The large scale employment of risk analysis is said to have supplanted other potential frameworks of mediating environmental, economic and other potential factors in food safety regulation, for instance the virtual obliteration of any mention of the precautionary principle in both SPS and Codex frameworks.[623] Another instance of this marginalization

---

[619]   Ibid.

[620]   David E. Winickoff and Douglas M. Bushey, 'Science and power in global food regulation: The rise of the Codex Alimentarius' (2010) 35 *Science, Technology and Human Values* 356, 364.

[621]   Ibid.

[622]   Ibid 360.

[623]   The terminological use of precaution was a major bone of contention for a decade during the formulation of the Principles and Guidelines for Microbiological Risk Management, resulting in a complete absence of the term in the final adoption. See: FAO/WHO, *Report of the 38th Session of the Codex*

regards the use of 'other legitimate factors' in the Codex framework. A general decision of Codex in 1995 had recognized that the Commission ought to regard 'other legitimate factors' relevant for the protection of consumers' health and fair practices in food trade, in appropriate circumstances.[624] Winickoff and Bushey describe the gradual obliteration of 'other legitimate factors' through the relentless and singular focus on standardization of risk analysis measures.[625]

Codex proceedings are said to rehearse and play out pre-existing sharp policy differences, in a manner similar to the experience in SPS and TBT Committees mentioned in the previous subsection. Livermore discusses several characteristics that make robust deliberation difficult within Codex, including large inequalities among participants, few procedural protections for minority positions, and over-representation of industry and state interests. The shift from the prior voluntary phase has aggravated the effects of these inequalities due to the significantly higher stakes in the WTO regime, including legal and economic (dis)incentives for non-compliance.[626] Poli identifies and elaborates how the differences between the EU and the US in three issues are played out in Codex, viz., the role of the precautionary principle in risk management decisions and the role of factors other than science in Codex decision-making, as well as the need to label and impose traceability requirements for food derived from biotechnology.[627]

Committee on Food Hygiene, Joint FAO/WHO Food Standards Program ALINORM 07/30/13; FAO/WHO, *Report of the 30th Session of the Joint FAO/WHO Codex Alimentarius Commission.* Rome: Joint FAO/WHO Food Standards Program ALINORM 07/30/REP.

[624] See further, *Codex Alimentarius Commission Procedural Manual* (16th edn, Codex Alimentarius Commission 2006).

[625] Winickoff and Bushey (n. 620) 361.

[626] Michael A. Livermore, 'Authority and legitimacy in global governance: Deliberation, institutional differentiation, and the Codex Alimentarius' (2006) 81 *New York University Law Review* 766, 769–771. However, the Appellate Body in *EC-Hormones I* had underlined the need to be prudent about the mandatory effect of such international expert standards, and to take a flexible approach of 'based on' these standards as opposed to 'confirm with' these standards (n. 426) para. 165. Further, the Appellate Body in *EC-Sardines* suggested these standards as a 'basis' or starting point rather than being the substantive end point (n. 617) para. 275. However, there is no reversal of burden of proof in such cases of failure to comply. See further: Lee, 'Ambiguity and hierarchy' (n. 159) 233.

[627] Poli (n. 618) 616–629.

Commentators note an increased emphasis on constructing scientific consensus through a conscious avoidance of any public acknowledgement of expert disagreement in the Codex, including an active mobilization of efforts by Codex to prevent votes in such cases.[628] Furthermore, voting has been reportedly avoided to such an extent that controversial issues have been left at the draft stage for a considerable number of years to avoid bringing them before a regulatory committee where the result would inevitably be a vote.[629] A publicized deliberation of differences in technical findings in the committee has been frowned upon, where the secretariat is reported to have insisted that 'you cannot vote in science; you can only disagree'.[630] The change in emphasis, here, is from recognition of disagreement in science and its reporting, to contriving to resolve such disagreement through institutionalized negotiation, which will strive to reach middle grounds, without an explicit recognition of the controversies inherent in reaching a traditional understanding of scientific consensus.[631]

The 'behind the scenes' generation of an expert consensus on standards by both Codex and the SPS Committee (described in the previous section) without necessarily resolving outstanding public issues is significant here; it becomes extremely crucial with the enforceability of these standards by the WTO Dispute Settlement Body.[632] Lang and Scott question the legitimacy of the managerialism of transnational experts, seen in the activities of these committees, dominating Members' regulation of health and the environment.[633] They discuss the implications of Weiler's work regarding the creation of a relatively close-knit community of trade negotiators and governmental officials within the GATT/WTO system, who share a defined 'ethos', a sense of common purpose, normative commitments, and common ways of defining and analysing

---

[628]   Lang and Scott (n. 530).

[629]   Ibid.

[630]   Winnickof and Bushey (n. 620) 370. They substantiated this concern with another illustration of voting. In the case of the Emmental Cheese label in 2007, a decision was apparently 'all but sealed', apart from a single dissent. However, when it was put through voting, a quarter of Members voted against the decision. This, the authors underlined, shows that in many cases of apparent consensus in technical decisions, a considerable number of experts might vote against, if a counter proposal is seconded.

[631]   Ibid 371.

[632]   Text to nn. 616–617.

[633]   Lang and Scott (n. 530) 601–605, fns 128, 138.

problems.[634] In this context they recall the concerns expressed about the side-lining of existing international institutions and states in such kind of technocratic governance by trans-governmental expert networks.[635] Domination of such expert networks is criticised as often 'not accountable to governments or parliaments back home'.[636] The concern of the democratic deficit, also seen in EU institutions, is amplified in expert networks that seek to position their mode of governance as a functional, problem-oriented process of deformalization, whereby the traditional political processes are hollowed out. In such circumstances, overt political contestations become subsumed in the discourse of expert cooperation, where 'struggles over global governance are to a certain extent ... fought through the debates waged within and between various scientific and professional disciplines and their universalizing discourses'.[637] It is conceivable that concerns of fragmentation can be resolved in a way that the transnational expert consensus can be arrived at, without attempting to address the concerns behind the divergent normative approaches to development and use of technologies that might fuel public contestations about specific technologies. Thus, even when systems of standards are in harmony with each other, if they are discordant with widespread citizen opposition in Member States, expert harmonization can only work up to a certain point in practice, and means very little in principle while addressing issues of political representation. The International Law Commission acknowledged this as such, when it recognized that fragmentation can often be caused by normative conflicts, expressions of diverging preferences and values rather than merely technical mistakes, and 'require[s] a legislative, not a legal technical response'.[638]

Taking stock, global rules that define the permissible ambit of EU rule-making appear to be dominated by a certain claim to scientific

---

[634] Cited in Joseph Weiler, 'The rule of lawyers and the ethos of diplomats: Reflections on the internal and external legitimacy of WTO dispute settlement' in Roger Porter, Pierre Sauve, Arvind Subramanian and Merico Beriglia Zampetti (eds), *Efficiency, Equity, Legitimacy: The Multilateral Trading System at the Millennium* (Brookings 2001) 334.

[635] Lang and Scott (n. 530) 604.

[636] Ibid. They cited the influential article of Julia Black, 'Constructing and contesting legitimacy and accountability in polycentric regulatory regimes' (2008) 2 *Regulation and Governance* 137.

[637] Sol Picciotto, 'Networks in international economic integration: Fragmented states and the dilemma of neo-liberalism' (1997) 17 *Northwestern Journal of International Law and Business* 1014.

[638] International Law Commission (n. 603) para. 484.

normativity. We have seen the supersession of the WTO regime over all others, and SPS over other Agreements within the trade regime, partly caused by the judgment of the Panel in *EC-Biotech*. It was also suggested that, in any case, a technocratic understanding dominates the implementation of the precautionary principle under the Cartagena Protocol. This flows from the discussion in Chapter 4 that if the trigger for the principle is held by techno-scientific communities, then public contestations about technological trajectories might not be appropriately represented in public decision-making.[639]

An insistence in the SPS regime on classical risk parameters, the problems of which were discussed in Chapter 3, is an inappropriate employment of scientific discourse since it is unrealistic in a context of emerging scientific knowledge and systemic incertitudes. Given the fact that in such situations, preliminary scientific studies that raise fundamental concerns about the assumptions of an existing scientific assessment cannot provide definite falsifying evidence, these closures through classical risk parameters are neither a reassurance that is respectful of public concerns, nor do they have a handle on how scientific discourses, enquiries and advances operate. Surveying the use of science during WTO dispute settlement, Herwig concluded that law can enhance its authority and epistemic validity through scientific evidence, but only if it recognizes limitations of science's epistemic and normative authority.[640] She emphasizes that law has to approach science as contested knowledge and risk regulation as political decision-making, leading inevitably to more indeterminate solutions to transnational legal conflicts. Connected to the discussion in Chapter 3, she observed:

> [I]n risk as a decision-making resource that enables decisions in the face of uncertainty, through a certain kind of probabilistic framing of hazards ... framing something as risk implies dealing with these uncertainties by either integrating them into the risk assessment where the sources of uncertainty are in principle known (for example, through stating the confidence level or error rate of a study) or blending them out of the assessment as too speculative and hypothetical pending the generation of further knowledge about these uncertainties.[641]

According to Herwig, when the notion of risk is primarily understood as a construction that is based on certain social and scientific conventions, and not as a fact, 'framing an issue as one of risk may sometimes

---

639   Text to nn. 447–465.
640   Herwig (n. 579).
641   Ibid 823.

coincide with there actually being a risk in the sense of a calculable probability of hazard. At other times, it may not'.[642] She continued, 'however, for all its emphasis on probability and the possibilities of taking decisions, a contrapuntal undercurrent of uncertainty remains when frames of risk are used'.[643]

Textual and interpretative readings of SPS provisions, including a strict reading of (in)sufficiency of scientific evidence, as well as the determination of the studies which can constitute a risk assessment and SPS measure, provide limited room for such a nuanced understanding of scientific uncertainty. As discussed in the previous chapter, scientific uncertainty is not necessarily exposed through expert disagreements, but rather public contestations about existing scientific certainty informed by their everyday experiences can lead to gradual recognition of those concerns by some experts, which may later lead to minority opinions and disagreements between experts.[644]

5.5 WINDOWS FOR WIDER REPRESENTATION IN THE SPS AGREEMENT

In a strictly formal sense, there is a window of opportunity to bring public opinion to the fore in the WTO legal process through submission of amicus briefs that may be put on record at the DSB, subject to judicial discretion.[645] Here, there is a limited amount of accommodation within WTO law in terms of Members' ability to be responsive to public opinion,[646] albeit one which is severely limited by the lack of clear procedures, as well as by the difficulties faced by non-parties in acquiring relevant information about the case.[647]

---

[642] Ibid 824.

[643] Ibid.

[644] Text to n. 400.

[645] *EC-Hormones I* (n. 426) para. 156.

[646] See Joanne Scott, 'European regulation of GMOs: Thinking about judicial review in the WTO' (2004) 57 *Current Legal Problems* 140; she terms Members' ability to be responsive as considerable, perhaps in reflection of the history of the lack of inclusion of non-state actors in transnational tribunals.

[647] See, Amicus Brief in *EC-Biotech* jointly filed by the Center for International Environmental Law, Friends of the Earth – United States, Defenders of Wildlife, Institute for Agriculture and Trade Policy and Organic Consumers Association – United States (1 June 2004) <http://www.ciel.org/Publications/WTOBiotech_AmicusCuriaeBrief_June04.pdf> last accessed 12 January 2012, fn 3.

The DSB, in disputes involving the provisions of the SPS, has only recognized the importance of considering public opinions and concerns at the stage of risk management, but not regarding how risk assessment is framed. Foster notes how the DSB has consistently interpreted that public appreciation of the magnitude of risk could be recognized in risk management, through a combination of Articles 3.3, 5.5 and 5.6 SPS (to set appropriate levels of protection in a manner that is neither arbitrary nor a disguised restriction on international trade).[648] However, important elements of risk assessment, viz., determining the sufficiency of available scientific evidence, the scientific principles appropriate in a technical enquiry, how to base measures on these principles, and how risk assessments and provisional measures are to be framed, are all held as processes in which public opinion is of no relevance. Foster argues that in cases of serious scientific uncertainty, a Member should be able to defend an SPS measure on the basis that its population simply does not want to undertake a particular risk. She emphasizes that though it may only be on the rarest of occasions that SPS measures are adopted with entirely pure motives, in those exceptional cases it is necessary to interpret the SPS Agreement in a way that will accommodate public concerns about an uncertain risk as a legitimate factor.[649] Given the technocratic understanding of the constitution of risk, as a product of magnitude and probability, she argues that there is a role for subjective assessment of magnitude through incorporation of public opinion, making public opinion an inherent part of risk assessment.[650] Thus public fears, if demonstrated to exist, can be a reason for not only the maintenance of higher levels of protection, but also inconsistencies in levels of protection between goods that are considered similar, apropos the WTO disciplines of rationality and proportionality. The acceptability of inconsistency in levels of protection between like-goods was suggested by the Appellate Body in *EC- Hormones I*, when it held that it would be

---

[648]   Caroline E. Foster, 'Public opinion and the interpretation of the WTO agreement on Sanitary and Phytosanitary measures' (2008) 11 *Journal of International Economic Law* 427, 431. Cf. Tracey Epps, 'Reconciling public opinion and WTO rules under the SPS Agreement' (2008) 7 *World Trade Review* 359, 383–385.

[649]   Foster, ibid 432.

[650]   This is in line with the submissions of five renowned social scientists in the second amicus brief to the WTO Panel in *EC-Biotech:* Amicus Curiae Brief, submitted to the Dispute Settlement Panel of the World Trade Organization in the case of EC: Measures Affecting the Approval and Marketing of Biotech Products (Lawrence Busch and others, 30 April 2004).

absurd to accept an insistence on similar levels of protection for beef developed with natural and synthetic hormones.[651]

The tension between international expert standards and appropriate levels of protection of a Member was identified in the previous section as opening legal spaces that could accommodate public contestations about technology. Public demands for higher levels of protection could pave the way for deviation from international expert standards, when these demands trigger further expert investigations and enquiry in risk analysis. The Appellate Body in *EC-Hormones I* appears to entertain possibilities of legality of concerns that are outside the strict sense of expert considerations in risk assessment. It exhorted the Panel 'to bear in mind that responsible, representative governments commonly act from perspectives of prudence and precaution where risks are irreversible, for example, life terminating, damage to human health are concerned'.[652] Though these remarks are contrary to the reality of the final findings of the Appellate Body,[653] it opens a window of opportunity for public estimations of risk that are different from those of techno-scientific communities to be taken into consideration during adjudication of the legality of Members' measures. This opportunity for wider publics to participate can include public perceptions regarding the magnitude of hazard and the linkages between such public perceptions to public preferences for certain risk frames over others. Such a possibility thus connects Members' textual right to choose their own level of protection to public opinion and political pressure. Thus one can identify a possibility, at least theoretical, of construing an opportunity to bring considerations of public concerns and opinion in risk inquisitions (at least in categories publicly considered to be serious and irreversible) even if an existing international expert consensus may have suggested otherwise. Pursuit of higher levels of protection can thus require scientific studies, and even identification of scientific ambiguity and ignorance.

Though textually all that is required by the trade regime is a rational relationship between findings of the risk analysis and the ultimate measure, the restricted ambit of which studies can constitute a risk

---

[651]   *EC-Hormones I* (n. 426) para. 187.

[652]   Ibid para. 187.

[653]   Ibid para. 253. It found fault with the specific minority opinion for a perceived lack of specificity, a mere suggestivity, and the lack of direct relationship to the identified risk. Joanne Scott's comment that the Appellate Body's bite was more muted than its bark is apposite, especially since there was a finding that the specific measure could not have been based on a risk assessment; Scott, 'European Regulation of GMOs' (n. 646) 139.

analysis in the *EC-Biotech* decision provides considerable challenges to the realization of this aforementioned window for Members' regulation to incorporate public contestations about expert characterizations. Related to this is the *EC-Biotech* Panel's problematic treatment of the term 'insufficiency of scientific evidence', that appears oblivious of differing conditions of insufficiency that correspond to different levels of protection (mentioned in an earlier section). This finding is however contradictory to the relevant interpretations of the Appellate Body in *EC-Hormones II*, which recognizes the necessity to conceptually link 'insufficiency of evidence' to 'levels of protection'. Coupled with the insistence that qualitative analysis would suffice as risk assessment in *EC-Hormones I* (mentioned in the previous two paragraphs), there is a slim chance for a meaningful window for wider public opinion in SPS frameworks, through recognizing public contestations about existing expert ascertainment, influencing identification of an appropriate level of protection.[654] However, for the realization of such a possibility, explicit, meaningful and institutionalized deliberative exercises during the framing of risk assessment are necessary. Through such exercises Members may be able to attempt regulatory innovations to find different means to coalesce expert and lay concerns in the characterization of an appropriate level of protection.

5.6 ROOM FOR PROTECTION OF PUBLIC MORALITY IN THE WTO REGIME

Conceptually separable, though intricately connected, is the avenue for protection of public morals during the regulation of GMOs within WTO rules on global trade. More than a decade ago, Pauwelyn identified a pertinent question about overlapping objectives of a Member's measures that is still relevant to us here:

> In all the three cases involving SPS [before *EC-Biotech*] the parties agreed it was a health measure, but what happens if the contention is that, say, the measure is only ten percent health, and the rest is consumer anxiety or protection of moral? ... or the same is not about protection of health at all? (sic.)[655]

---

[654] Such a hope is also noticed in Emily Reid and Jenny Steele, 'Free trade: What is it good for? Globalization, deregulation, and public opinion' (2009) 36 *Journal of Law and Society* 11.

[655] Pauwelyn (n. 565) 644.

Deviations from the principles of non-discrimination for protection of public morals, provided in GATT XX (a), are part of general exceptions that allow Members to use measures that violate GATT obligations in order to pursue certain national policy objectives.[656] Marwell has identified a number of reasons for a heightened attention to the trade–morality interface in recent years, including an increased cultural heterogeneity in the membership of the WTO and the heightened 'disciplining' of environmental, health and other related regulation by the Appellate Body.[657] He identified an absence of judicial elaboration as a contributing factor for Members to recast regulations that might fall foul of the SPS Agreement under the public morals exception. The trade–morality interface has been examined by a trade tribunal in only two cases (*US-Gambling* and *China-Audiovisuals*) in the last fifty years.[658] The Panel in *EC-Biotech* recognized that a single measure can be considered and justified under different Agreements so that despite violations of the SPS Agreement as a health measure, it could still be justified under GATT Article XX (a) as a measure to protect public morals.[659] Such a position is again recognized by the Panel and Appellate Body in *China-Audiovisuals*, where the public morals exception was found to be a possible defence against violations to China's *Accession Protocol* to the WTO.[660]

The WTO legality of measures that is claimed to fall under the exceptions in Article XX is generally evaluated through a two-tier test, and the Member issuing the specific measure must demonstrate that it is

---

[656]   Text to nn. 532–542.

[657]   Jeremy C. Marwell, 'Trade and morality: The WTO public morals exception after Gambling' (2006) 81 *New York University Law Review* 802, 808–811. Appendices at the end of this article list examples of regional and bilateral free trade agreements that incorporate public morals clauses, as well as countries that maintain morals-related trade-restrictive measures.

[658]   WTO Panel Report *United States – Measures Affecting the Cross-Border Supply of Gambling and Betting Services* (*US-Gambling*) WT/DS285/R, adopted 10 November 2004, and WTO Appellate Body Report WT/DS285/AB/R, adopted 20 April 2005; WTO Panel Report *China – Measures Affecting Trade Rights and Distribution Services for Certain Publications and Audiovisual Entertainment Products* (*China-Audiovisual*) WT/DS363/R, adopted 20 March 2009, and WTO Appellate Body Report WT/DS363/AB/R, adopted 19 January 2010.

[659]   *EC-Biotech* (n. 562) Add. 1 to Add. 9, and Corr.1, DSR 2006: III–VIII, 847.

[660]   Panel Report, *China-Audiovisual* (n. 658) para. 4.112 and Appellate Body Report *China-Audiovisual* (n. 658) para. 23.

necessary. Important elements of this judicial evaluation were identified in the second section of this chapter. It included a demonstration of how the impugned measure was designed to pursue a policy objective within the 'defined scope' of the relevant exception and a 'necessity test' to prove the required degree of connection between the measure and its objective. Subsequently, a good faith requirement in the *chapeau* ought to be met by demonstrating that the measure neither arbitrarily discriminates between like conditions/products, nor is a disguised restriction on international trade.[661] Diebold suggests that each of the elements mentioned below is to be met separately for the judicial validation of a measure under these general exceptions:

(i)   Is the measure aimed at a policy objective other than a trade restriction?

(ii)   Does the objective of the policy at issue fall within the scope of one of the exhaustively listed public interests (protection of public morals in this case)?

(iii)   Does trade in the good or service at issue actually pose a risk to or impairment of the achievement of the public interest or moral sought?

(iv)   Does the connection between the measure and the policy objective meet the required level?

(v)   Whether the measure either arbitrarily or unjustifiably discriminates between countries where like conditions prevail or constitutes a disguised restriction on trade (good faith requirements of the *chapeau*)?[662]

The definition of public morals is a threshold enquiry for the applicability of this defence since Members need to demonstrate that specific measures fall within this definition of public morals. The Panel in *US-Gambling* employed an ordinary meaning of public morals as denoting 'standards of right and wrong conduct maintained by or on behalf of a community or nation'.[663] This dispute arose out of a complaint from

---

[661]   See Christoph T. Feddersen, 'Focusing on substantive law in international economic relations: The public morals of GATT's Article XX (a) and "conventional" rules of interpretation' (1998) 7 *Minnesota Journal of Global Trade* 75.

[662]   See for more: Nicolas F. Diebold, 'The Morals and Order exceptions in WTO law: Balancing the toothless tiger and the undermining mole' (2007) 11 *Journal of International Economic Law* 43, 47–48.

[663]   Panel Report *US-Gambling* (n. 658) para. 6.465. Though this was in the context of the GATS Article XIV (a) exception it is similar in scope regarding

Antigua regarding US measures to curb Antigua-based online gambling services. The respondent invoked the defence of protection of public morals, by asserting that social harms from organized crime, money laundering and underage gambling necessitated restrictions on internet gambling to protect public morality. The Panel recognized that Members had some scope to define and apply the concept of public morals in their territories, in accordance with their own systems and scales of values, given that the content of public morals 'can vary in time and space for members depending upon a range of factors, including prevailing social, cultural, ethical, and religious values'.[664] Given the scenario that an internationally uniform definition of morals may not correspond to local values, protection of national morals needs to be delinked from any attempt to find an internationally uniform definition of morals. Such an approach is also palpable in the morality exception under Article 27 (2) TRIPS, public morals exception to the protection of freedom of speech under Article 10 (2) of the ECHR, and the position of UNCTAD and ICTSD that Members have considerable flexibility to define the situations where morality exceptions can be covered, depending upon their conception of public values.[665] Both the Panel and the Appellate Body in

textual treatment of public morals. Charnovitz concluded that the method of ordinary meaning employed to interpret the public moral exception under GATT by the tribunals, in light of the measure's object and purpose, only leads to ambiguous results: Steve Charnovitz, 'The moral exception in trade policy' (1998) 38 *Virginia Journal of International Law* 689, 701, 716. Diebold, in his review of authorities on the subject, found that the employment of Article 31, Vienna Convention in terms of its ordinary meaning, context and object and purpose, was quickly found to be dismissible, since it provides no useful results in the interpretation of public morals. He emphasized it to be 'at best difficult, but likely impossible'. Diebold (n. 662) fn 17.

[664] *US-Gambling* (n. 658) para. 6.461. However, the Panel found that US measures were not necessary to vindicate public morality, which was subsequently overturned by the Appellate Body. Further, the Appellate Body found that despite establishment of necessity, the US failed to show that its laws did not discriminate against foreign gambling services. See for a succinct account of the facts of the dispute, including the lack of compliance by the respondent: Sayera J. Iqbal Qasim, 'Collective action in the WTO: A "developing movement" toward free trade' (2008) 39 *University of Memphis Law Review* 153, 159–167.

[665] UNCTAD/ICTSD, *Resource Book on TRIPS and Development* (Cambridge University Press 2005) 379, cited in Diebold (n. 662) fn 58. He also discusses the experience of the public policy and order public exceptions in private international law, 56–57. Cf. Mark Wu, 'Free trade and the protection of public morals: An analysis of the newly emerging Public Morals Clause Doctrine', (2008) 33 *Yale Journal of International Law* 215. He opined that

China-Audiovisuals (the other case in which public morals was used in a WTO tribunal) 'assumed' that all the asserted measures were driven by the need to protect public morals, given the absence of any protest from the complainant on this count.[666]

The Appellate Body in *US-Gambling* upheld the Panel's employment of the definition of public morals, and noted that depending on the facts and circumstances of the case the evidence of existence of public morals may be derived from international practice, national ordre public, and public policy and national legislative history.[667] Given the dynamic interpretation of the notion of public morals in *US-Gambling*,[668] Marwell persuasively argued that Members should be able to define public morals based solely on their internal circumstances, pursuant to certain evidentiary requirements, since public morals are 'highly subjective, geographically localized, and diverse across political boundaries'.[669] The legality of specific regulatory measures in EU GMO regulation, through the exception for protection of public values, depends upon the nature of public morals that is invoked in the DSB, as well as its purpose and connection to the impugned measure. Lee reflected on whether concerns about animal welfare, interfering with nature, consumers' right to know and the distributional impacts of regulation could fall within the ambit of public morals here.[670] She opined that while it is unlikely that the Appellate Body will allow the provision to turn into a 'catch-all' provision (given the available defences under Article XX (b) and (g) GATT), it is not totally inconceivable that 'a combination of argument, information on context, and evidence of public views could be used to demonstrate that specific measures aim at protection of public morals'.[671]

---

various issues need further clarity, including whether the measure asserted as a measure of public morals *requires* universal backing through reference to international state practice or endorsement by relevant international organizations, and whether the defence would be judicially viewed differently between *inward* and *outward* directed measures (231–236).

[666] Further, the Appellate Body in *China-Audiovisuals* observed that (in the context of another assumption of the Panel that it deemed unacceptable) assumptions may not always provide a solid foundation upon which to rest legal conclusion, detract from a clear enunciation of the relevant WTO law and create difficulties for implementation (n. 658). 213.

[667] Appellate Body Report *US-Gambling* (n. 658) paras 296, 299.

[668] Panel Report *US-Gambling* (n. 658) para. 6.461.

[669] Marwell (n. 657) 824–826, 836.

[670] Lee, 'EU Regulation' (n. 15) 195.

[671] Ibid.

Significantly, whether the regulatory objective of safety is pursued within the ambit of public values of a Member is an important facet of this general exception. Reactions of regulators to systemic incertitudes and scientific ignorance can differ between Members in accordance with their respective public morals and values. These values include the manner in which regulators treat the instability of underlying knowledge in the regulation of GMOs and the institutional ethics of risk regulation.[672] Davis succinctly argues that public values that influence these factors of regulation are within the subject matter of the general exception of protection of public morals under GATT.[673] As many sections of this book have suggested, there appears to be general disaffection in the EU towards the sole reliance on expert constructions of safety through classical risk parameters, hinting at a particular moral position regarding the collective manner in which society reacts to systemic incertitudes.[674]

Given a wide and permissive definition of public morals in *US-Gambling*, the centrality of the WTO tribunal's role through evaluations of necessity and 'weighing and balancing' becomes stark. Standards and means of evidence become important to ascertain questions like whether trade in a specific good actually impinges protection of the relevant public morals and if the impugned measure is the least trade-disrupting policy alternative available to ensure the specific protection. Any judicial scrutiny as to whether an outright ban of a specific transgenic plant is necessary to protect public morals depends on the facts and circumstances of the case. It is discernible that questions about the least restrictive measure and non-discrimination between like-goods would be intricately connected not only to the application of a measure that is purported to protect public morals, but even to the nature of the

---

[672] Text to nn. 780–800.

[673] Gareth Davies, 'Morality clauses and decision making in situations of scientific uncertainty: The case of GMOs' (2007) 6 *World Trade Review* 249. Cf. Brownsword (n. 24) 204: 'where there is a level of scientific uncertainty that leaves room for legitimate disagreement, those states that prefer to take a risk-averse approach are allowed, at least provisionally to do so; on the other hand, where there is little room for scientific doubt, members are not to be encouraged to dress up their moral objections as if they are concerns about safety. Objecting that one does not want to gamble on GM crop safety is one thing; objecting against GM crops ... on moral grounds is something else'.

[674] In conventional terms the range of these options is characterized as permissive, prudential, precautionary or restrictive. See further, Adam D. Sheingate, 'Promotion versus precaution: The evaluation of biotechnology policy in the United States' (2006) 36 *British Journal of Political Science* 243.

morals that are invoked. Lest excessive application and abuse should transform it to a Trojan horse for the WTO system, the means of evidence to determine the objective and application of any policy measure as a moral exception becomes important.[675] Though in *US-Gambling* morality is broadly viewed according to a Member's own systems and scales of value, it is conceivable that eventually Members will have to provide concrete evidence that the specific measure pursues the objective of protecting domestic public morals. Further, it may need demonstration that the specific regulatory measure is in good faith in terms of the graveness of the effect or significant number of people affected; as also regarding the necessity of the impugned measure, as opposed to a less trade-restrictive measure or narrower application.[676] Diebold argues that a majority of a Member's population must endorse a particular value for this provision to be applicable.[677] He compared his position to the proposal of Pascal Lamy (as Commissioner of DG Trade) for a safeguard clause related to the clash between trade and collective preferences.[678] Lamy advocated a position where 'certain collective choices are binding on a society as a whole and transcend individual preference' and are undertaken through an internal review (through widespread consultations) of the collective preference regarding the measure adopted, in order to establish whether it was well founded.[679] As suggested in the first chapter, it is arguable using statistics, surveys and various heuristic devices that a significant number of people in the EU are sceptical of the large-scale use of GMOs in agriculture. Though numbers could become crucial in a judicial examination, the means of ascertaining any view as majoritarian can be challenging in cases where there are deep divisions through multiple axes of moral preferences.

The earlier sections noted a narrow window of possibility in the SPS regime to accommodate public participation in risk assessment and management; a regime that currently dominates the other Agreements in

---

[675] 'Effective *ex ante* mechanisms for controlling moral hazard often entail effective means of monitoring the behaviour of contracting parties *ex post*': Robert Howse, 'China – Measures affecting the protection and enforcement of intellectual property rights' (2011) 10 *World Trade Review* 87.

[676] Appellate Body Report *US-Gambling* (n. 658) paras 300–302, 300–311, 317.

[677] Diebold (n. 662) 66–67.

[678] Ibid.

[679] Pascal Lamy, 'The emergence of collective preferences in international trade: Implications for regulating globalisation' (15 September 2004) <http://ec.europa.eu/archives/commission_1999_2004/lamy/speeches_articles/spla242_en.htm> last accessed 12 January 2012.

the risk regulation of GMOs through judicial construction. The current section noted the right of WTO Members to initiate regulatory measures that seek to protect public morality. How the regulation of research, development and use of GMOs falls within the ambit of such protection is subject to context and argument. We also noted that this context can include contrasting public values regarding how regulators treat systemic incertitudes and scientific ignorance in the regulation of new technology. Since such values could be subjective, it becomes important to understand how the public is represented and allowed to participate in the ascertainment of public values about GMOs in the EU. The national opt-out clause from GM agriculture, proposed by the Commission in 2010, expects Member States to identify the 'social, economic and ethical aspects' in a manner that increases the level of public involvement. A referendum to gauge public opinion for collective choices could be part of ascertaining a claim on public values. However, establishing the existence of public opinion *simpliciter*, for or against a policy measure, may not be conceptually sufficient to establish public values. How public values about GMOs are currently mediated in EU regulation, and whether it includes legal measures to ensure participation of publics wider than techno-scientific communities, need further elaboration. This is attempted in the following chapter.

6. Public contestations and pursuit of public values

> Bioethical debate will have to become more political, and take fuller
> cognizance of the realities of the contemporary world, its technologies
> and its institutional possibilities in order to deal with the ethical
> fragility of individualist conceptions of informed consent.[680]

This chapter investigates the room for wider public participation in the pursuit of public values during EU regulation of the development and use of GMOs. Competing and conflicting values exist not only about consumption but also about the development of technologies like GMOs in the EU. The implications of not focusing on the existence of conflicting public values are to allow the values dominant among the industry and techno-scientific communities to have a free rein in the design and development of technology. If development and use of technology have to be justified in terms of commensurate public values, the regulatory routes to identifying and elaborating those values become crucial. Hence it is important that law facilitates the characterization and representation of these public values during the regulation of GMOs. The description here continues from the discussion in the previous chapter, where the need for elaboration of the manner in which public values are pursued in the EU GMO regulation was noted.

Currently, there are two strategies apparent in the EU regulation of GMOs in its engagement with protection of public values. The Irish Council for Bioethics succinctly identified the first, viz., protection of the public from harm by progressing in a cautious and stepwise manner (through risk regulation), and protecting the ethical autonomy of the consumer to choose through adequate labelling and coexistence strategies.[681] Labelling laws help pursue personal ethical choices of consumers

---

680   Onora O'Neill, 'Informed consent and genetic information' (2001) 32 *Studies in History and Philosophy of Biological and Biomedical Sciences* 689, 702.

681   Irish Council of Bioethics, *Genetically Modified Crops and Food: Threat or Opportunity for Ireland? Opinion* (Irish Council of Bioethics 2005) 36

inclined to consume GM products, while others averse to them can avoid them. The implications of this labelling framework that seeks an ethically neutral structure in law's engagement with public values about research, development and use of GMOs will be examined and elaborated in the next section. It recalls the discussion about the general requirement to label GMOs in Chapter 2, and further describes requirements for supplementary ethics labelling in situations where ethical or religious concerns about specific GMOs are identified. The section notes the contradictory and multi-interpretable nature of consumer concerns, and underlines the need for further clarity as to which substantive ethical concerns of a consumer can conceivably be included under the provision for supplementary ethics labels. It further emphasizes the inappropriateness of a technical authority like EFSA to make judgements about these normative aspects.

Subsequently, the section examines the vision of coexistence measures, for creating an ethically neutral platform of segregation through which all forms of agriculture can flourish, as an integral part of consumer and producer choice. It identifies the conceptual and practical controversies in the coexistence framework that are incompatible with this vision of ethical plurality. These controversies include the Commission's insistence on coexistence as a purely economic issue, the difficulties in ascertaining and apportioning the burden of putting in place a segregation structure, and doubts about the technical (im)possibilities of a foolproof segregation mechanism. They raise fear in some quarters that coexistence facilitates large scale GM cultivation, and creates a fait accompli of large scale inadvertent mixing of crops. The section further emphasizes and elaborates the limitations of the vision of an ethically neutral framework through which ethical plurality can flourish.

Apart from the attempt at creating a neutral framework that respects the ethical plurality of consumers, the regulatory regime also uses advice from public bioethics committees to identify, reflect and represent issues of public values related to the regulation of science and technology.[682] The Deliberate Release Directive underlines the importance of respect for ethical principles recognized in Member States and, in accordance with the principle of subsidiarity, emphasizes the right of Member States to

---

<http://www.bioethics.ie/uploads/docs/GM%20Report1.pdf> last accessed 21 December 2011.

[682] Prominent advisory bodies on ethical questions relating to science and new technologies include GAEIB (Group of Advisers on the Ethical Implications of Biotechnology), its successor the EGE and various National Bioethics Committees.

consider ethical aspects during the deliberate release of GMOs.[683] The Directive envisages consultations with the Commission's EGE, for obtaining advice on ethical issues of a general nature regarding the deliberate release or placing on the market of GMOs, either *suo motu* by the Commission, or at the request of the European Parliament, Council or a Member State.[684] It is emphasized that such consultations with this expert group on ethics should be without prejudice to the competence of Member States regarding ethical issues.[685] Provision for consultation of the EGE, or any other appropriate body, to obtain advice on ethical issues regarding placing GM food or feed on the market is also mandated by the Food and Feed Regulation.[686] Given the general claim by EU regulators of public participation as a way of representation noted in the previous chapters, the remainder of the chapter identifies the room for public participation in the pursuit of related public values opened up by public bioethics committees during the current EU regulatory regime of GMOs.

Towards a sufficient understanding of this second strategy a subsequent section highlights the supposed ethical turn in regulation, a turn that seeks to move beyond the reliance on ethical assumptions of the techno-scientific community about its own research. It further elaborates the conceptual employment of public bioethics in technology regulation, and regards the generation of public bioethics advice as part of the public sphere. Thus it underlines the public purpose of public bioethics to identify, characterize and represent the full range of values and conflicts regarding regulation of technology. It further identifies concerns about the impediment of such public functions during the working of expert-dominated committees, when they emphasize promotion of technology as an imperative starting point of ethical investigation, whereby an important opportunity for a public scrutiny regarding choices about trajectories of technological development is lost.

A small number of reports from the EGE and Member State committees have examined ethical issues arising from GMOs in agriculture under the current regulatory regime. In two subsequent sections of this chapter, the advice of three public GMO-ethics committees that has been published during the current EU regulatory regime is examined. The attempt here is to analyse their respective approaches regarding the characterization of public contestations about technology, identify the important recommendations in these three reports, and the openings these

---

683 Recital 8 of the Deliberate Release Dir.
684 Recital 57 ibid.
685 Article 29.1 ibid.
686 Recital 42 of the Food and Feed Reg.

might provide for wider public participation in EU regulation of GMOs. The advice of the Irish Council for Bioethics, the first of the three reports, is found to be indicative of the criticisms of a conventional approach to GMO-ethics visible in earlier public bioethics committees. The approach of the report is examined in the light of these criticisms, and focuses on the little room it leaves for any other ethical concerns outside the heuristic of techno-scientific risk. In the subsequent section, the reports of the EGE and the Danish Council of Ethics are examined, underlining the positive manner in which earlier criticisms about GMO-ethics are taken on board. It identifies the implications of these reports for an appropriate characterization and a full blown civic deliberation of public contestations and concerns. In the final section, a specific species of recommendations in the EGE and Danish Council reports, viz., environmental impact assessment, utility assessment and technology assessment, is highlighted for its potential for facilitating wider deliberation of regulatory values about GMOs. This focus on impact assessment in the section is not intended as a suggested policy solution to implement public participation. It is merely to highlight the possibilities of wider participation in regulation that technology assessment exercises can potentially offer, as opposed to a strict techno-scientific conceptualization of risk regulation. On the whole, the evaluation of the second strategy of public bioethics committees in EU regulation of GMOs has two elements. First, whether the generation of such advice includes a process of *lay* participation. Secondly, irrespective of the composition of public bioethics committees, whether the ethical scrutiny and recommendations in these reports provide room for participation wider than expert communities. Altogether, the chapter seeks to identify the normative assumptions behind reliance on these two aforementioned strategies, and how public participation and public contestations about technology are accommodated within them.

6.1 PURSUIT OF ETHICAL PLURALITY THROUGH LABELLING AND COEXISTENCE MECHANISMS

A prominent strategy for the pursuit of public values in EU regulation of GMOs is through the recognition of ethical pluralism, facilitated by labelling laws and coexistence strategies to ensure segregation between GM crops and other crops. The legislative objective of labelling as a measure to inform and communicate with consumers is understood by ethics committees as an integral part of the citizen's right to know and to make informed choices.[687] Since consumers may have competing and conflicting ethical positions about consuming GMOs, this strategy assumes that the appropriate way forward would be to create an effective labelling and segregation mechanism through which a plurality of values of consumers and social groups can coexist in a market. This is seen as an important strategy within a classical liberal model to ensure individual autonomy, since the state is expected not to force particular perceptions of the good life on individuals, allowing them to determine their own conceptualizations about such values in situations of disagreements.[688] Pursuit of consumer autonomy within EU regulation of GMOs is facilitated through a general obligation to label GMOs (elaborated in Chapter 2), and through supplementary labelling mechanisms mandated for specific cases where ethical or religious concerns are conceivable (elaborated later in this section).

The Preamble to the Food and Feed Regulation provides:

Labelling should include objective information to the effect that a food or feed consists of, contains or is produced from GMOs. Clear labelling, *irrespective of the detectability* of DNA or protein resulting from the genetic modification in the final product, *meets the demands expressed in numerous surveys by a large majority of consumers, facilitates informed choice* and precludes potential misleading of consumers as regards methods of manufacture or production. In addition, the labelling should give information about any characteristic or property which renders a food or feed different from its conventional counterpart with respect to composition, nutritional value or nutritional effects, intended use of the food or feed and health implications for

---

[687] Group of Advisers on the Ethical Implications of Biotechnology, *Opinion 5 on Ethical Aspects of the Labelling of Foods Derived from Modern Biotechnology* (GAEIB 1995) <http://ec.europa.eu/bepa/european-group-ethics/docs/opinion5_en.pdf> last accessed 21 December 2011.

[688] See for an elaboration of this liberal model of autonomy and ethical plurality: Danish Council of Ethics, *Utility, Ethics and Belief in Connection with the Release of Genetically Modified Plants* (Danish Council of Ethics 2007) 98–99. Further, three important conditions for a consumer's choice to be autonomous are identified, viz., competence (psychological and physical capacities to make autonomous choices), authenticity in desires and beliefs (freedom from coercion and constraint) and power to implement these beliefs/desires to choices. The existence of alternative choices is found to be necessary for the exercise of an individual's power to implement his or her beliefs, that is, a choice of refusing to buy a particular kind of food cannot be meaningful if the only alternate choice to avoidance is having to grow the food. Helen Siipi and Susanne Uusitalo, 'Consumer autonomy and availability of genetically modified food' (2011) 24 *Journal of Agricultural and Environmental Ethics* 147.

certain sections of the population, *as well as any characteristic or property which gives rise to ethical or religious concerns.*[689]

These objectives of the Regulation give fuller expression to the more general statements of intent found in the objectives of the Deliberate Release Directive that labelling ought to ensure appropriate identification.[690] These statements are substantiated through the general requirements to label GM products in the Directive and the Regulation.[691]

6.1.1 Supplementary Labels Stipulated for Ethical Concerns

The Food and Feed Regulation also stipulates an additional label in situations where any characteristic or property of a GMO gives rise to ethical or religious concerns. The trigger for these supplementary ethics labels is tied to an earlier stage of the regulatory procedure, viz., the application for authorization to place any GM food or feed on the market. Under the Regulation, applications for authorization are expected to be accompanied by a reasoned statement that the specific product does not give rise to ethical or religious concerns; alternatively the application should be accompanied by plans for supplementary labels in accordance with the provisions in the Regulation.[692]

Supplementary ethics labelling procedures can be initiated in a straightforward manner if applicants propose a specialized labelling scheme on their own accord. However, it is unclear how the regulatory system can ascertain the veracity of a 'reasoned statement' in instances where an application avers that a specific ethics label is not required due to the absence of any ethical or religious concerns. From the discussion in Chapter 2, the reader may remember that the application for authorization is to be assessed by EFSA, including the requirements for supplementary labels.[693] Here, the technical safety Authority is left with little guidance (not to mention legitimacy or competence) to decide on

---

[689] Recitals 21, 22 of the Food and Feed Reg. Emphasis added.

[690] Recital 40 of the Deliberate Release Dir.

[691] Article 26 and Annex IV of the Deliberate Release Dir. sets a framework for implementation of labelling requirements through committee mechanisms. Articles 12, 24, 13 (1) and 25 (1) of the Food and Feed Reg. substantiate them. See for a more detailed description, text to nn.139, 217–223.

[692] Articles 5 (3) (g) and 17 (3) (g) Food and Feed Reg.

[693] Though 'with assistance from an appropriate food assessment body for safety assessment, or an appropriate competent authority for environmental risk assessment, or a community reference laboratory for methods of detection and identification' under Article 6 (3) of the Food and Feed Reg.

the particulars of a reasoned statement of the applicant that a specific transgenic product does not give rise to ethical or religious concerns. This factor brings out an element of uncertainty to the institution of mandatory labelling that provides supplementary information for variegated ethical concerns. The impact of this lacuna on how public participation is furthered during the protection of public values depends on the nature of values to which consumers can legitimately expect these ethics labels to cater. It is therefore necessary to elaborate further on the range of concerns that consumers may pursue through a label.

The importance of not conflating consumer concerns and citizen values (amply emphasized in the work of Mark Sagoff)[694] is relevant here, since it might be inappropriate and unwise to consider the market, and consumer ethics, as a more secure route to efficacious expression of public values than other public channels. However, if it is to be used as one of the prominent routes, it is important that consumers receive meaningful and accurate information that is reliable and verifiable; although what information is meaningful in this context needs further elaboration.[695] In a number of studies, consumer demands for mandatory labelling were found to be not merely expressions of lack of trust in safety evaluations of environment and health characteristics of a product.[696] They were also found to include an opportunity to register their view about the wisdom of food biotechnology –in other words, 'to support or not support the dissemination of food biotechnology, apart from their views about the characteristics of the individual product'.[697] A

---

[694]   Mark Sagoff, *The Economy of the Earth* (Cambridge University Press 1988) 7–10.

[695]   See for a technical proposal for rewarding food suppliers for seriously taking ethical concerns of consumers into account: Tassos Michalopoulos, Michiel Korthals and Henk Hogeveen, 'Trading "ethical preferences" in the market: Outline of a politically liberal framework for the ethical characterization of food' (2008) 21 *Journal of Agricultural and Environmental Ethics* 3. The proposal pertains to the development of the ethical dimension of food production through a labelling mechanism, and focuses on ethical characterization of what information is meaningful and most important for this endeavour.

[696]   US FDA Center for Food Safety and Applied Nutrition, *Report on Consumer Focus Groups on Biotechnology* (Alan S. Levy and Brenda M. Derby eds, Office of Scientific Analysis and Support 2000), cited in Douglas Kysar, 'Preferences for processes: The process/product distinction and the regulation of consumer choice' (2004) 118 *Harvard Law Review* 526, fn 361. Further, Recital 2 of the Food and Feed Reg. cites a similar sentiment among 'numerous surveys by a large majority of consumers'.

[697]   Ibid (US FDA).

general label that informs the process of genetic modification within a product might be able to cater to this kind of straightforward concern.

However, Korthals pointed out that several types of consumer concerns about agriculture and food are contradictory and multi-interpretable.[698] This underlines the dynamic nature of pluralistic consumer concerns, since they are responding to complex socio-technological situations and developments. He described how ethical concerns of food consumers in Europe often concern both substantive issues and issues relating to transparency and alienation. First, structural traits of the food chain have become important substantive ethical concerns of food consumers.[699] For GMOs, these include concerns regarding the acceptable level of intervention in nature,[700] accentuation of distributional issues and increased corporate control of food chains through GM seeds,[701] and the increase in intensive and commercialized agriculture at the expense of de-intensified and re-localized models associated with peasant or family farming.[702] Secondly, ethical concerns related to transparency pertain to the reliability of information related to food choices, for example, 'that consumers with a preference for organic meat products look for different information about the food chain and want different advice than consumers with other preferences'.[703] Thirdly, ethical concerns of alienation pertain to the widespread consumer feeling of disconnection and lack of control in the way food is produced in food chains.[704]

---

[698] Michiel Korthals, 'Ethical rooms for maneuver and their prospects *vis-à-vis* the current ethical food policies in Europe' (2008) 21 *Journal of Agricultural and Environmental Ethics* 249.

[699] Ibid 258.

[700] The level of intervention in GMOs is seen as qualitatively higher than the conventional breeding mechanisms due to the far higher degree of control of agro-ecological conditions in GM agriculture; ibid.

[701] Sometimes, this accentuation is through the technological possibilities offered by GM technology, which aids stricter control for implementation of patent provisions. See for an elaboration Danish Council (n. 688) 146–147.

[702] Rosa Binimelis, 'Coexistence of plants and coexistence of framers: Is an individual choice possible?' (2008) 21 *Journal of Agricultural and Environmental Ethics* 437, 439.

[703] Korthals (n. 698) 258.

[704] Ibid: Some consumers simply take this gap for granted, while others may find it troubling and try finding out where their food comes from, 'very often with disappointing results, because they do not get a satisfying answer to their query or cannot get any information at all'. He emphasizes the fundamental connections between the ethical concerns of involvement and participation with the earlier-mentioned substantive concerns.

Over and above a general GM label, it is unclear whether the aforementioned range of contradictory and multi-interpretable ethical concerns legally requires supplementary labels for each specific ethical concern. However, if consumers have to meaningfully participate in decisions of public values during their consumption, then this range of information needs to be provided on the label. If public contestations about GM technology are indeed arising from different intellectual, political and ethical concerns, then the earlier identified lacuna regarding the stipulation of supplementary ethics labels through EFSA becomes significant. This is since a supplementary ethics label can mandate the inclusion of information regarding the technological impact of the product on different ethical concerns, for instance about distributional issues or accentuation of corporate control, that might have different implications in different GM products. Though there is no clarity on the substantive contents and limits of ethics in this supplementary labelling provision, nor any sociological investigation available, the competence of EFSA to make such judgements is suspect.

6.1.2 Coexistence Measures

Stringent upstream segregation of seeds and crops becomes crucial for an effective labelling system in agricultural products so as to communicate accurate, verifiable and reliable information. Thus coexistence measures that ensure effective segregation (elaborated in Chapter 2) are instrumental for realizing consumer autonomy, as well as representing and facilitating a plurality of ethical choices. This subsection examines the difficulties regarding realization of the regulatory strategy of protecting public values by pursuing ethical pluralism and consumer autonomy, which are sharply visible in the controversies around coexistence measures.

In an area that is traditionally understood to be largely a matter of Member State authority, the Commission's beginning point for implementing coexistence is found in its recommendation that 'no form of agriculture, be it conventional, organic, or agriculture using GMOs, *should* be excluded in the EU'.[705] The vision of coexistence as an important platform for ensuring ethical plurality is, however, at sharp

---

[705] Commission Recommendations, 2003 (n. 228) Recital 1. Article 26a of the Deliberate Release Dir. mentions coexistence as a matter of Member State responsibility, but see text to nn. 228–230 regarding how this subsidiarity authority is sought to be hemmed in by the 'non-binding recommendations' of the Commission. Further, for the constraints on Member State flexibility through

variance with the strictly economic approach emphasized by the Commission in its first set of recommendations.[706] Lee criticises the singular emphasis in the Commission's guidelines on coexistence as an economic problem, which negates its importance for the effectuation of ethical pluralism.[707] Since 'the narrow economic understanding of coexistence … implies that to the extent that presence of GM material' has marginal market impact, the labelling provision will then 'raise no coexistence issues'.[708] The Commission's insistence that coexistence measures are purely economic is inconsistent with an aspiration for ethical plurality through labelling and consumer choice. This is because 'meaningful coexistence is absolutely central to the consumer choice rhetoric … since in the absence of distinctiveness between the three forms of farming, consumers cannot choose between them'.[709] In contrast, the Committee of Regions stressed the intrinsic rather than economic value of different forms of agriculture, which is far more consonant to an approach based on ethical plurality.[710] Representations of such intrinsic essence can be found, for instance, in the Regulation on Organic Production, which recognizes the role of organic agriculture not only to cater to consumer demand, but also to deliver an important public good beyond the market, like 'contributing to protection of the environment and animal welfare, as well as to rural development'.[711]

As discussed in Chapter 2, the current set of Commission recommendations has continued to emphasize that an economic approach is central to EU coexistence measures.[712] Limitations arising from the Commission's insistence on a strictly economic approach to coexistence make aspirations of an ethically neutral framework less coherent. The Commission's insistence impacts the contours of a 'neutral framework', which is especially visible with regard to the question of the type of agriculture

---

functioning of COEX-NET, see Lee, 'Governance of Coexistence' (n. 225) 196–198.

[706]   Commission Recommendations, 2003 (n. 228) Recitals 4, 5.

[707]   Lee, 'Governance of Coexistence' (n. 225).

[708]   Ibid 199.

[709]   Ibid.

[710]   Committee of Regions, 'Opinion of the Committee of the Regions on the Communication from the Commission to the Council and the European Parliament report on the implementation of national measures on the coexistence of Genetically Modified crops with conventional and organic farming' (2007) OJ C 57/11.

[711]   Organic Reg. (n. 225) Recital 1.

[712]   See n. 232.

(conventional, organic or GM) that bears the cost of segregation.[713] The Commission guidelines required farmers introducing a new production type to bear the responsibility for implementation of segregation strategies.[714] However, Lee reminds us that the daily grind of ensuring coexistence falls on those parties (organic or conventional farmers) that have an interest or compulsion to avoid GM material, and not vice versa.[715] Such efforts include institution of controlling weeds, changing planting regimes, and isolating equipment to ensure effective segregation of their crop and produce. In contrast, GM produce is not economically affected by the presence of organic or conventional produce.[716] Apart from the risk of loss of market value for organic or conventional produce, non-GM farmers may have to label their produce as containing GMOs in case of contamination; they also have to anticipate the burden of avoiding 'innocent infringement' of patent rules, in situations of inadvertent mixing with a neighbour's GM crop.[717] Notably, the current set of Recommendations recognizes this additional burden to the organic farmer

---

[713] Commission Recommendations, 2003 (n. 228) para. 2.1.7. With respect to a preference by Member States to impose costs on GM farmers, whether or not they were first to farm an area, see: Commission Staff Working Document, 'Annex to the Report on the implementation of national measures on the Coexistence of GM crops with conventional and organic farming' SEC (2006) 313,16, cited in Lee, 'Governance of Coexistence' (n. 225) fn 68.

[714] Ibid (Commission Recommendations, 2003) para. 2.1.7.

[715] Lee, 'Governance of Coexistence' (n. 225) 209.

[716] Article 9 Organic Reg. (n. 225), which stipulates that any farm produce that has to be labelled as containing GMOs cannot be labelled as organic.

[717] See the application by organic farmers for certification as a class, regarding a suit for compensation due to the spread of GM canola in Canada, a spread which made it virtually impossible for organic farmers to grow canola: *Larry Hoffman, LN Hoffman Farms Inc and Dale Beaudoin* v *Monsanto Canada and Bayer Cropscience Inc* 283 DLR (4th) 190 [2005] SKQB 225 (First Instance). The Court did not examine the reasonableness of the assertion of harm to the applicants, but rejected the application on a technical ground that a class action suit is not maintainable under the relevant statute, since the Court saw the damage as purely economic, which precluded the applicants from being considered as a class. However, in *R.* v *Secretary of State for the Environment and MAFF, ex parte Watson* [1999] *Environmental Law Review* 310, an organic farmer in the UK unsuccessfully sued his neighbour for cross-pollination from field trials of GMOs. The Court of Appeal decided that while field pollination cannot be guaranteed, the government had arrived at a perfectly reasonable point at which to strike the balance between the competing interests in play. See also: *Karl Heinz Bablok* (n. 145), where the Court reaffirmed that honey containing traces of pollen from GM plants must be labelled according to the general procedure prescribed in the Food and Feed Regulation. See further, Maria Lee

from a farm-by-farm approach of coexistence underlined by the 2003 Recommendation: 'in certain cases, and depending on market demand and the respective provisions of national legislation ... presence of traces of GMOs in particular food crops – even at a level below 0.9% – may cause financial harm to the operator who would wish to market them as not containing GMOs'.[718]

Ethnographic studies among rural farmers in Catalonia highlight local tensions in farming communities (between farmers who do not want to plant GM crops and those who do) due to the individualization of ethical and other responsibilities through coexistence measures.[719] This tension is visible in the case of the creation of GM crop-free zones in various parts of the EU, and a network of GMO-free regions, whereby various local farmer groups and communities (as well as some regional and local governments) seek the effectuation of their respective ethical worldviews. They perceive GM crops as a threat to other crop productions and ecological sensitivities that are rooted in an ethical model of post-productivism or alternative agro-developmental model.[720] The Commission recommendations recognize that there may be economic and natural conditions where Member States could consider the possibility to exclude GMO cultivation. It insists that such exclusion should rest on the demonstration that other measures are not enough to achieve sufficient levels of purity.[721]

Further, the very idea of a legitimate and economically viable coexistence framework is found doubtful by many groups. These doubts relate to differences in the normative emphasis and basis of coexistence measures; that is, if they ought to be restricted to an economic choice model, or they should explicitly recognize other ethical concerns as well. Doubts persist about ecological and technical (im)possibilities of crop segregation. This includes the identification of appropriate refuge between GM and other crops and realistic thresholds to ensure effective

---

and Robert Burrell, 'Liability for the escape of GM seeds: Pursuing the "victim"?' 65 *Modern Law Review* 517.

[718] Commission Coexistence Recommendations, 2010 (n. 231) para. 1.1. However, the Court in *Pioneer (Italy)* (n. 228) remarked that the current set of Recommendations simply describe and amplify the 2003 Recommendations, albeit without offering any explanation or elaboration.

[719] Binimelis (n. 702). See also, the examination of the assertions of impossibility of coexistence by the Irish Council for Bioethics (n. 681) 44–45.

[720] Yann Devos and others, 'Ethics in the societal debate on genetically modified organisms: A (re)quest for *sense* and *sensibility*' (2008) 21 *Journal of Agricultural and Environmental Ethics* 29, 42–43.

[721] Commission Coexistence Recommendations, 2010 (n. 231) Annex 2.4.

prevention of gene flow from one type of agriculture to another. The European Environmental Agency conducted a study that showed the incidence of pollen-mediated gene flow in all the major relevant crops (maize, rape, sugar beet) to be very high, and suggested that the isolation distances (through barrier crops) required are too high to work.[722] In contrast, another study by the EC Joint Research Centre in the same year concluded that coexistence was feasible with adjustments in current farm practices.[723] Binimelis reviewed a series of authors who have highlighted the impossibility of coexistence between organic and GM-based agriculture, due to the incommensurability of rationale at the technical and conceptual level.[724] The difficulties in ensuring fool-proof segregation, as also to live with a certain amount of mixing between the three forms of agriculture, have brought to focus the controversies of the GM thresholds in organic and conventional products. The recent Commission Recommendations reveal that zero level of mixing is not even a goal for regulators, implicitly accepting that a certain amount of admixture will be a reality, and thus assuming that regulating levels of mixing are only important if there is an economic impact.[725]

The controversies surrounding the acceptable levels of adventitious admixture are also related to fears that these would serve as a back-door push for large-scale introduction of GM crops.[726] Thus the notion of coexistence measures in support of consumer autonomy and ethical pluralism is fraught with a number of conceptual and practical inadequacies. These inadequacies make hollow a vision of public participation in

---

[722] Katie Eastham and Jeremy Sweet, 'Genetically modified organisms: The significance of gene flow through pollen transfer' (Environmental issue report No. 28, European Environment Agency 2002) <http://www.eea.europa.eu/publications/environmental_issue_report_2002_28> accessed 14 October 2011.

[723] European Commission, *Scenario for Co-existence of Genetically Modified, Conventional and Organic Crops in European Agriculture: A Synthesis Report* (Joint Research Centre 2002).

[724] Binimelis (n. 702) 259. See further: Laura Ponti, 'Transgenic crops and sustainable agriculture in the European context' (2005) 25 *Bulletin of Science Technology Society* 289; Levidow and Boschert, 'Coexistence or contradiction?' (n. 236); Kathleen McAfee, 'Beyond techno-science: Transgenic maize in the fight over Mexico's future' (2008) 39 *Geoforum* 148.

[725] Commission Coexistence Recommendations, 2010 (n. 231) Annex 2.3.2: 'Member States should consider that there may be no need to pursue specific levels of admixture where labelling a crop as GM has no economic implications'.

[726] See for concerns about how large scale cultivation of canola and rape in Canada, soy and maize in the US, and soy in Argentina, ended up with difficulties in producing non-GM crops of the same species; n. 63.

pursuit of public values during use of GMOs through consumer choice. In fact, attempts to elaborate coexistence measures have brought back unresolved conflicts about values and ideals regarding GM crops, rather than effectuate ethical plurality through institutional neutrality. They demonstrate the inherent limitations of a strategy to pursue ethical pluralism and consumer autonomy to resolve the clash of public values in GMO regulation, and diminish possibilities of meaningful representation in GMO regulation through coexistence and consumer information.

6.1.3   Other Conceptual Limitations of a Strategic Reliance on Consumer Autonomy

The strategy of pursuing consumer autonomy and ethical pluralism by making available all types of products may not be able to trump ethical concerns like the protection of producer autonomy or the autonomy of suppliers/processors/wholesalers/retailers. However, some of these concerns are in conflict with each other. For instance, farmers cannot be forced to produce a kind of product for the purpose of making available those kinds of products in pursuit of consumer autonomy.

Indeed, even assuming full realization of autonomy for consumers and producers, there may be other overriding public values in the regulation of technology that could take precedence. Such concerns, including environmental values, other people's well-being, beneficence and social justice, may trump an approach of individualized ethics and consumer autonomy. Therefore, this necessitates institutional interventions beyond facilitation of consumption choices.[727] Ethical issues about wealth can be relevant to techno-scientific research. For instance, ethical considerations regarding the collection, research and use of plant biodiversity that culminate in GM crops include ensuring adequate recognition of the contributions of indigenous communities and traditional farmers to that process.[728] It is doubtful if citizens who consider these ethical concerns

---

[727]   See further, Siipi and Uusitalo (n. 688).

[728]   Gripping points in existing codes of conduct for such property issues about agro-biodiversity research in general, or of a particular company in particular, could be ethical principles like equity, the inalienability of rights of local communities and farmers towards active participation in the research, full disclosure regarding the extent and ultimate purpose of the research to these communities, their prior informed consent, ensuring integrity of the morality and spirituality of the culture, traditions and relationships of indigenous peoples, appropriate restitution of indigenous peoples should any adverse consequences occur from research activities, and active support of indigenous research:

as values of general importance will be satisfied with merely avoiding the consumption of these GM products, as opposed to influencing the ethics of the development of such technologies themselves. If that is the case, then a vision of an ethically neutral framework through which ethical plurality flourishes is an inadequate strategy to mediate contestations about public values.

The construction of a labelling and coexistence mechanism aimed at ensuring ethical plurality through the facilitation of consumer choice to prefer or to avoid transgenic products can have some degree of virtue in law's engagement with public contestations about technology, since it provides the consuming publics arenas to participate in the regulatory process. However, the aforementioned inherent limitations of this approach accentuate the importance of pursuing more participative routes of engaging with public values in the regulation of research, development and use of technologies like GMOs.

6.2 PURSUIT OF PUBLIC VALUES AND PUBLIC BIOETHICS

The second prominent EU strategy identified in the regulation of GMOs is the use of bioethics for deliberating whether the regulation of research, development and use of GMOs falls within acceptable markers of public values. Callahan identifies the endeavour of bioethics as catering to a need for a more open dialogue between science and society towards ascertaining, 'so far as that is possible, of what is right and wrong, good and bad, about the scientific developments and technological deployments'.[729]

Use of ethics as a separate component of regulatory decision-making in science and technology anticipates the inadequacies in unduly relying on the techno-scientific community in the pursuit to make research, development and use of technology compatible with corresponding public values. Some commentators link the growing trend of constitution of 'a wide range of commission-like advisory bodies to assure legitimacy of political action in ethical questions' to law's need to demonstrate the legitimacy of regulatory actions, despite intense political divisions and

---

Johannes M.M. Engels, Hannes Dempewolf and Victoria Henson-Appollonio, 'Ethical considerations in agro-biodiversity research, collecting, and use' (2011) 24 *Journal of Agricultural Environmental Ethics* 107.

[729] Daniel Callahan, 'Why America accepted bioethics' (1993) 23 *Hastings Centre Report* 8–9.

destabilization of the instrument of classical risk.[730] Here, ethics is seen as an important and decisive semantic form in which governance discourses are conducted in situations where political opposition is not framed in terms of left/right but in terms of right/wrong.[731] The relevance of public bioethics in this context is the moving away from an assumption of a monopolistic reliance on the ethical viewpoints generated in the techno-scientific community, and the creation of spaces and processes to speak about the development and use of new technologies in a structured and democratic way.[732]

The Warnock Committee process in the UK is an early marker of law's need to address the dilemmas of conflicting public values brought forth during research, development and use of new technologies, like GMOs. It examined the controversies about the use of *in-vitro* fertilization (IVF) in the UK. The Warnock Report recognized this fundamental dilemma of liberalism, viz., of securing freedom (of science and the individual) without endangering the foundations of social order, or 'how to protect the foundations of social order without illegitimately curtailing the freedom of the individual'.[733] In relation to the need to elicit ethical

---

[730]   Alexander Bogner and Wolfgang Menz, 'How politics deals with expert dissent: The case of ethics councils' (2010) *35 Science, Technology and Human Values 888*, 894. A standard text traces the emergence of bioethics as a 'term with social currency' to the early 1970s, when academics, predominantly in the field of medical ethics, philosophy and law, became engaged in the study of 'the moral dimension – including moral vision, decisions, conduct and polities'– of the lifesciences and ... health care, employing a variety of ethical methodologies in an interdisciplinary setting: Albert R. Jonsen, *The Birth of Bioethics* (Oxford University Press 1998) 3. Salter and Jones emphasized the growing importance of the field of bioethics by using crude quantitative indices like the number of courses, conferences, websites and committees, and discussed its growing political utility due to the growth in its advisory role in regulatory frameworks: Brian Salter and Mavis Jones, 'Biobanks and bioethics: The politics of legitimation' (2005) 12 *Journal of European Public Policy* 710. Dodds and Thomson cite the institution of over ninety National Bioethics Committees around the globe as a marker of the public importance attached to bioethical issues: Susan Dodds and Colin Thomson, 'Bioethics and democracy: Competing roles of national bio-ethics organisations' (2006) 20 *Bioethics* 326.

[731]   See further, Mouffe (n. 314).

[732]   Alfred Moore, 'Public bioethics and public engagement: The politics of proper talk' (2010) 19 *Public Understanding of Science* 197, 202.

[733]   Ibid. See further: UK HMSO, *Report of the Committee of Inquiry into Human Fertilization and Embryology* (UK HMSO 1984) <http://www.hfea.gov.uk/docs/Warnock_Report_of_theCommittee_of_Inquiry_into_Human_Fertilisation_and_Embryology_1984.pdf> accessed 14 September 2011.

advice, Warnock observed that a society without 'inhibiting limits, especially in the areas with which we have been concerned (IVF), would be a society without moral scruple ... what is common is that people generally want some principles or other to govern the development and use of the new techniques'.[734]

Biotechnology was the area in which ethics as a component in EU decision-making in science and technology was first envisaged through the formation of GAEIB; all the other areas of science and technology were later brought within the ambit of GAEIB.[735] GAEIB's successor, the EGE, recognized the European Charter of Fundamental Rights, which emphasizes that the Union is founded on the indivisible and universal values of human dignity, freedom, equality and solidarity as well as the principles of democracy and the rule of law, as part of the broad ethical framework in the EU for the research, development and use of science and technology.[736]

As mentioned earlier in the chapter, the Deliberate Release Directive empowers the Commission to consult any committee (such as the EGE) to obtain advice on the ethical implications of biotechnology, either *suo motu* or at the request of the European Parliament, Council or a Member State.[737] The proposal for including 'social, economic and ethical aspects' to provide the platform for Member State decisions on cultivation of EU authorized GMOs (under the Directive) was highlighted in the second chapter. This proposal has the potential to provide a participatory window to articulate public values, if and when further steps are taken to codify the proposal in the statute book.[738] The Food and Feed Regulation also provides for the Commission to consult the EGE, or any other appropriate body established by the Commission, for an opinion on

---

[734] Mary Warnock, *A Question of Life: The Warnock Report on Human Fertilization and Embryology* (Blackwell 1985) 2.

[735] European Commission, 'Commission Decision on the renewal of the mandate of the European Group on Ethics in Science and New Technologies' (11 May 2005) 2005/383/EC.

[736] EGE opinion (n. 116) 35–36. See further: *Charter of Fundamental Rights of the European Union, 2000,* 2000/C 364/01 <http://www.europarl.europa.eu/charter/pdf/ text_en.pdf> last accessed 24 November 2011. The EGE opinion also recognized the importance of the UNESCO *Universal Declaration on Bioethics and Human Rights,* 2005, including the principles of human dignity, consent, autonomy and responsibility, privacy, equity and justice, solidarity and benefit-sharing.

[737] Article 29 of the Deliberate Release Dir.

[738] Commission Communication (n. 141).

ethical issues on its own initiative or at the request of a Member State.[739] Given the advisory nature of these committees, how their recommendations may be implemented in the decision-making structure is opaque and indeterminate. However, the Directive emphasizes that consultation with any such committee ought to be conducted under 'clear rules of openness, transparency and public accessibility', and the outcome of such consultation be made accessible to the public.[740] Since ethical issues are largely a matter for Member States, ethics committees of a number of Member States have given advice on various aspects of regulation of GMOs, including the UK Nuffield Commission and the National Councils of Ireland and Denmark.

6.2.1 Conceptual Employment of Public Bioethics in Regulation

Before we examine the room made available for public participation through such advice, it is necessary to examine how public bioethics is conceptually employed in regulation for understanding its implication for the problem of representation in regulation of technologies like GMOs. This is a difficult task in view of the 'absence of empirical work on its institutions, ideologies, knowledges and notions of ethical expertise as part of the difficulty of gauging the precise nature of the identity of bioethics as moral philosophy'.[741] Moore describes public bioethics as a complex of institutions, practices and discourses connecting policy-making with ethical considerations and ethical deliberation, for improving political decision-making.[742] He points out how public bioethics is only one element of the broader regulatory ethics (differentiating it from academic bioethics, corporate bioethics or bioethics in corporate institutional locations, as well as the work of clinical and research ethics committees), and that public bioethics committees have a 'direct or

---

[739] Article 33 of the Food and Feed Reg. The textual differences between the Deliberate Release Dir. and the Regulation are significant, including (a) the Parliament and Council are not capable of directly requesting the EGE and so on, for an opinion under the Regulation, and (b) there is no explicit requirement in the Regulation that the consultations of the EGE and so on, be conducted in a manner that is open, transparent and publicly accessible.

[740] Article 29.2 Deliberate Release Dir.

[741] Salter and Jones (n. 730). They noted the difficulty of gauging the precise nature of the identity of bioethics as moral philosophy, or as driven by utilitarian service of interests rather than a search for truth.

[742] Moore (n. 732) 198.

indirect connection to the state, that is, in one way or another initiated, supported, controlled or commissioned by state actors'.[743]

When an academic endeavour conducted by experts to describe what is right and wrong, or good or bad, becomes involved in formation of public policy, 'moral thought has to be radically reinvented and reframed' for its use as a regulatory tool to characterize public values.[744] Such reframing includes emphasizing the inevitable connection of public bioethics to the public sphere, and less to methodological justifications of analytical moral philosophy. This is important because public bioethics needs further legitimation than a claim of competence in analytical philosophy and ethical reasoning.[745] This is since such advice can influence setting of policies, which will be generally applied to all people in a polity, in the name of values of all.[746] Dodds and Thomson identify various responsibilities of National Bioethics Organizations (NBOs), including 'contributing to and stimulating public debate, providing expert opinion in identifying relevant issues that need to be addressed in policy deliberation, and/or developing public policy'.[747] In contentious ethical debates where there is a need to reflect diversity of opinion and ethical frameworks, they argue for NBOs to open up public debate by a well-informed response and amplification of both expert and lay public views. The appropriate identification and an amplified description of all positions and interests are important during disagreements on public

---

[743]   Ibid.

[744]   Mariachiara Tellichini, 'Governing by values. EU ethics: Soft tools, hard effects' (2009) 47 *Minerva* 281, 282.

[745]   John H. Evans, 'Between technocracy and democratic legitimation: A proposed compromise position for common morality public ethics' (2006) 31 *Journal of Medicine and Philosophy* 213, 214. He follows Charles Taylor in his conception of public sphere as 'a common space in which the members of society are deemed to meet through a variety of media: print, electronic, and also face-to-face encounters; to discuss matters of common interest; and thus to be able to form a common mind about these': Charles Taylor, 'Liberal politics and the public sphere' in Amitai Etzioni (ed.), *The New Communitarian Thinking* (University Press of Virginia 1995) 183, 185.

[746]   Mairi Levitt, 'Public consultation in bioethics. What's the point of asking the public when they have neither scientific nor ethical expertise?' (2003) 11 *Health Care Analysis* 15.

[747]   Dodds and Thomson (n. 730) 327. The authors moot a basic framework to develop public policy involving NBOs which seek to promote well-informed and diverse expression of opinion in society. Here they identify building internal capacities to meet the democratic ideal of effective participation of citizens as key.

values about the development and use of technology. Without participation and involvement of a wide range of publics towards this end, the danger is that ethics would become a direct expression of political views of the state, and an illegitimate and undemocratic regulatory tool for dominant regulatory groups 'heavily reducing ethical enquiries and discussion to an expression of the intellectual establishment'.[748]

6.2.2 Conceptual Errors of Expert Domination in Public Bioethics

The call for wider public participation, arising from the concern that values of dominant regulatory groups may pass off as public bioethics, is usually accompanied by the problem of the domination of experts in the composition of public bioethics committees. This is since their composition can substantially influence proceedings and outcomes of public bioethics committees.[749] A study found that professionals from medical genetics and law constituted 62 percent of the members of public ethics committees on bio-banks across the world, hinting at the elevation of expertise in law and biological sciences as the most appropriate qualification to make ethical judgements about the interests and values of citizens in another important emerging area of bioethics.[750]

Such expert-centred generation of bioethics advice is in stark opposition to a vision of public bioethics committees that are composed from a wide range of publics, articulated in the UK Warnock Report. Following

---

[748]   Tellichini (n. 744) 282.

[749]   Bogner and Menz discussed different models of composition of public bioethics committees, the implications of those which do not include explicitly identified stakeholders and different sections of the public, and the implications of advice of various counter-ethics commissions and citizens ethics committees that arrive at substantially different findings from the official committees, while adhering to established practices of the bioethics community (n. 730) 894–897.

[750]   Salter and Jones (n. 730) 728. Another study showed a similar finding regarding the predominance of medical science and law in ethical committees: Jean-Christophe Galloux, Arne Thing Mortensen, Suzanne de Cheveigne, Agnes Allansdottir, Aigli Chatjouli and George Sakellaris, 'The institutions of bioethics' in Martin Bauer and G. Gaskell (eds), *Biotechnology: The Making of a Global Controversy* (Science Museum 2002) 129. Salter and Jones also point out that philosophers engaged in broader approaches to bioethical interrogation as a primary occupational pursuit 'are present in numbers unlikely to be sufficient to challenge (should they choose to do so) the assumptions of their scientific and legal colleagues'. Further, they note the addition of 'representative of the citizen interest' as 'a new breed of bioethicist coming from a variety of backgrounds including professions such as nursing, social science and policy', and interest groups like patients' organizations and disabled groups; ibid.

six years of intense public debate on IVF (both inside and outside Parliament), the Committee recommended permission for research on human embryos with explicit restrictions through licensing and other regulation, obliquely leading to the provisions in the UK Human Fertilisation and Embryology Act, 1990. Warnock criticized Devlin's influential conceptualization of a societal consensus on public morality in *Enforcement of Morals*, where Devlin had famously argued that if there is a consensus of public opinion against a certain practice, then only then must the law intervene to prevent that conduct.[751] Warnock criticized this assumption of common morality, and argued that the idea of a consensus in public opinion about such a morality is a myth:

> [T]here is no agreed set of principles which everyone, or the majority, or any representative person, believes to be absolutely binding, and this is especially so in areas of moral concern which are radically and genuinely new. The question must be recast: in situations where people disagree with each other as to the rights and wrongs of a specific form of behaviour, how do we decide whether or not the law is to intervene?[752]

Importantly, Warnock argued that the primary role of such committees was not to provide correct answers to the question 'what should we do?', but instead to concern themselves with framing a process in which decisions could be made despite profound disagreements. The committee recommended the constitution of a model of bioethics committees with significant non-scientific lay representation (greater than 50 per cent), for counterbalancing the professional interests of the scientific and medical community, and underlined the danger of possible exploitation by enthusiastic scientists if they are left with a dominant voice.[753] It is significant that the EGE and most Member State bioethics committees have no such specific provisions to ensure lay membership. Warnock is raising fundamental questions here about both the rationality and general functioning of public bioethics committees, as to whether members of the public are not capable of establishing their own frames of meaning. Such frames might differ from analytical reasoning of technical and scientific experts, including lawyers, theologians and moral philosophers. If that is the case, the difficulties of representing moral concerns of at least some lay publics by expert committees are significant. This aspect is elaborated in the subsequent subsection.

---

[751] Patrick Devlin, *The Enforcement of Morals: Maccabaean Lecture in Jurisprudence of the British Academy* (Oxford University Press 1959).
[752] Warnock (n. 734) xiv.
[753] Warnock Report (n. 733).

The Commission's Expert Group on Science and Governance raised fundamental questions about how ethics can be appropriately viewed as just a matter of expertise, and still convincingly represent 'the values of all Europeans' when it is institutionalized through the creation of expert committees.[754] Such use of ethics is found by the report to 'largely reduce to the bureaucratic mechanism of expert ethical advice, deriving from procedures which are identical to those for scientific advisory committees'.[755] It is found to move away from 'an emergent pluralist, inclusive and inter-disciplinary dialogue' as potentially a new way of shaping public policy around science and new technological endeavours.[756] The report asserts that if 'the need to establish a more intense and open dialogue between science and society' requires to 'ground the normally self-referential ethics of science in a democratically legitimated framework of values and socio-technical choices', then this reduction to bureaucratic mechanism of expert ethical advice is inappropriate.[757] This is since 'ethics is represented as if it is naturally *only* a matter of expert judgment, though this very framing has markedly shaped, and continues to shape, which ethics and whose values count in European politics'.[758] Here the slippery slope in normative language from 'European legislation on ethics' to 'ethical legislation' is seen to facilitate an undeliberated state production and imposition of (state values as) collective values.[759] The criticism here is that ethics through expert committees is in practice institutionalized as soft law without collective debate.[760] It thus neutralizes political issues, introduces norms outside traditional process of law-making by evoking society without involving it, and pays lip-service to democratic concerns while only expert processes take place; and thereby pertinent issues of representation are ignored.[761]

---

[754] Expert Group on Science and Governance (n. 53) 52.
[755] Ibid.
[756] Ibid.
[757] Ibid 46.
[758] Ibid.
[759] Ibid 47.
[760] Ibid.
[761] See further, *European Parliament resolution on institutional and legal implications of the use of 'soft law' instruments*, 4 September 2007. 2007/2028(INI), P6TA (2007) 0366.

6.2.3   An Imperative for Promotion of Technology through Consensus

Earlier in the section it was posited that the public function of public bioethics is intricately connected to the public sphere, where public debate is stimulated by the identification of the full range of values concerning contestations about the development of technology by a public body. Characterization of a range of values and their respective rationalities to the fullest possible extent is in direct contrast to an a priori emphasis on promotion of specific technologies seen among many public bioethics committees. Such an emphasis on promotion can be seen in the title of the Communication from the Commission to the Parliament and Council, which advocated accommodating ethics within EU decision-making on biotechnology, viz., *Promoting the Competitive Environment for the Industrial Activities based on Biotechnology within the Community*.[762] The emphasis on promotion (as opposed to a more neutral investigation on various public values about biotech research, development and use) is visible in the Commission's association of ethics with avoidance of an uncertain and confused debate, as 'such confusion can adversely influence the whole climate for industrial development of biotechnology'.[763] Significantly, the Commission ruled out proposals for a broad social assessment of new biotechnology through public discussion, on the basis of the 'imperative to avoid a situation creating uncertainty [which] could result in a diversion of investment and could act as a disincentive for innovation and technological development by industry'.[764] Such an a priori position during the generation of bioethics advice negates the aspiration for better citizen representation in technology regulation.

It is significant that although the European Parliament initially supported the institution of GAEIB, it subsequently passed a resolution noting the exclusion of parliamentary involvement (including in GAEIB's composition), and criticized it for paying too much attention 'to the interests of research and not enough to the possible effects on society'.[765] In this context, Salter and Jones remarked:

---

[762]   European Commission, 'Promoting the competitive environment for the industrial activities based on biotechnology within the community' Communication to Parliament and the Council SEC 629 final.

[763]   Ibid 11.

[764]   Ibid 8.

[765]   European Parliament, *Resolution on the mandate of the group of advisers on the ethical implications of biotechnology to the European Commission 1997* B4-0484/97. See further, Tellichini (n. 744) 288–289.

[I]t is generally the case that while institutionalized ethics of this nature may include public concerns within the policy process, they subordinate those concerns to the dominant political culture ... governmental ethics regimes are found to work as a form of reflexive government in which inclusion, involvement, and mobilization of extra-scientific actors, and perspectives are built into a discursive and institutional framework that stabilizes rather than destabilizes the commitment to techno-scientific progress and economic competitiveness.[766]

The Commission's Life Sciences strategy emphasized how public sensitivities require a change in the form of regulation of biotechnology to strengthen its legitimacy through a stated policy of enhanced public consultation and transparency in the functioning of regulatory committees, including a broadened definition of the expertise required for membership in regulatory committees.[767] Salter and Jones identified this stated shift within the context of a basic regulatory dilemma of attempting to tap the scientific and industrial potential to be a global leader in new biotechnologies on the one hand, and the acknowledgement within regulators that public support for such a project is essential, but presently scarce.[768]

If ethical scrutiny through public bioethics committees is to be an important form of eliciting public opinion, then an a priori position of promotion cannot be consistent with democratic principles. This is since the regulator's normative acceptance of the development and promotion of particular technological trajectories itself is a subject of public concerns, requiring further public scrutiny in ethics committees. A general aspiration for consensus towards promoting a particular technology in a public bioethics committee can be inconsistent with democratic principles in situations where public contestations about choosing such technological trajectories are strong, and where no shared conviction of an a priori acceptance of promotion is reasonably conceivable within society. As an analogy, a consensus recorded in any ethics committee report on the production of stem cell lines, stating a shared

---

[766]   Salter and Jones (n. 730) 713.

[767]   European Commission, 'Life Sciences strategy' (n. 91). Rather than a more open government that can readily politicize previously uninvolved constituencies, Salter and Jones speculated that a more attractive option for regulators to pursue scientific and industrial potential to be a global leader in new biotech is the use of a 'fresh authoritative source of expert advice, clearly different from that of science but one which can be integrated with the regulatory process using the same organizational form of expert consultation' (n. 730) 713.

[768]   Ibid (Salter and Jones) 713.

conviction that such production is legitimate for reasons of contributions to scientific progress and therapeutic uses, would be impossible 'if not for the fact that representatives of the churches, or from philosophical positions beyond the utilitarian mainstream' do not find a voice in the committees. Thus committees ought to be required to ethically scrutinize a priori positions of promotion itself, so as to facilitate a public deliberation about values regarding the development of GMOs.

Bogner and Menz argue that expert dissent ought to be seen as a normal feature of committee process that can conceptually enhance the salience and legitimacy of politics, since decisions on ethical issues cannot be taken on the basis of expert knowledge alone in any case.[769] Given the inevitability of serious divergence within public bioethics committees (if they are representative of a range of moral worldviews), it requires bioethics committees to record dissent and disagreement within a society. This concern requires both the record of dissent within its members and the seeking out of rationales behind disagreement within society so as to represent and provide full rationality to all conflicting points of view. Wider membership and participation here would be key to amplify law's ability to democratically and effectively engage with public contestations about technology. Though lay participation cannot be seen as a panacea that will automatically ensure a range of views and the protection of minority opinions, a full characterization of public consultations and votes on divergent recommendations are necessary complements to ensure an appropriate representation of a range of public values. Such practices can be seen in the functioning of the Danish Council of Ethics, examined in a later section of this chapter.

Having placed public bioethics advice as part of the public sphere, and emphasized its public function to identify, characterize and represent the full range of values and conflicts concerned with the regulation of technology, the next two sections examine the advice of three public GMO-ethics committees published during the current EU regulatory regime. The attempt here is to identify the significant recommendations in these three reports, analyse their respective approaches regarding characterization of public contestations about technology, and see if their ethical scrutiny amplifies room for public participation and better representation in the EU regulation of GMOs. The next section examines the advice of the Irish Council for Bioethics, and identifies criticisms against a conventional approach to GMO-ethics visible in earlier public bioethics committees. The section finds the Irish advice as symptomatic of these

---

[769]   Bogner and Menz (n. 730) 899.

criticisms, which is related to the seamless interpretation of public concerns as private feelings, having profound implications in terms of a narrower scope for ethical scrutiny. Following this, the subsequent section examines the relevant reports of the EGE and the Danish Council of Ethics, and underlines the positive manner in which previous criticisms about GMO-ethics are taken on board. It identifies the implications of these reports for an appropriate characterization and a full blown civic deliberation of public contestations and concerns about GMOs.

6.3 IRISH COUNCIL ADVICE: TYPICAL LIMITATIONS OF TRADITIONAL GMO-ETHICS

The Irish Council for Bioethics established a working group on GMOs in response to a request from the Irish Food Safety Authority in 2002. It considered the ethical issues raised by adopting GM crop and food technology in general, and in the context of Ireland in particular. This process culminated in a final report in 2005, and importantly included public consultations through prepared questionnaires.[770] The report held that genetic modification of crops is not morally objectionable by itself, since on balance the technology held a great deal of promise despite introducing new risks for consumers, farmers and the environment.[771] It held that an appropriate ethical approach to regulation would be to protect the public from harm by progressing in a cautious and stepwise manner (through risk regulation), and in the meantime to protect the autonomy of the consumer to choose through adequate labelling and coexistence strategies. In other words, when there is a choice for all parties, the potential benefits of GM crops can be made accessible to those who wish to avail of them.[772] This prescription of an appropriate framework on ethics of GMO regulation in the report is fairly similar to the regulatory approach that is apparent in the current EU release framework. The report substantially relied on the statement that:

> [T]he scientific community has widely agreed that the risks for human health associated with GM crop consumption are very low given the thorough safety assessments required for market approval. The magnitude of the environmental risks is more difficult to estimate, however ... these risks appear to be manageable through careful implementation strategies, and therefore the

---

[770] Irish Council (n. 681).
[771] Ibid 36.
[772] Ibid.

introduction of GM crops into Irish farming would not necessarily be irreversible.[773]

It also noted that:

> [A] scientific risk assessment involves predicting possible harmful conse-
> quences and estimating their likelihood, and such successful prediction of the
> consequences of any new technology is extremely difficult, and unanticipated
> side effects commonly emerge from the introduction of new technologies. The
> risks reviewed above should not cause panic; they reflect the inherent nature
> of scientific progress, which will always involve some side effects.[774]

The unequivocal belief in the scientific community and epistemology of science without ethical scrutiny and elaboration, based on a normative aspiration of progress, is noteworthy. This is notable due to the fact that significant public contestations arise from doubts about this unequivocal reliance on scientific risk as the strategy for regulation of release of GMOs.

How can a public bioethics report aspire to characterize and represent public contestations about the exclusive reliance on scientific experts, when it merely adopts such reliance in its report without any critical elaboration? Opinions of the majority of respondents to the questionnaire prepared by the working group of the report contradicted this position of reliance, which in turn required some reflection and further ethical elaboration from the Council about this position. An overwhelming majority stated that they do not trust scientists or governmental organ-izations to provide all the relevant and available facts.[775] For this reason, they stated that they are not convinced of the safety of GM foods that were authorized for sale. They expressed their concerns about possible adverse impacts on future generations, and were not confident that GM food/crops are carefully regulated, including the effectiveness of coexist-ence mechanisms to segregate GM agriculture from organic agriculture. A large majority of respondents thus did not support controlled cultiva-tion of GM crops in Ireland.[776] While the report appreciated 'the vigour and sincerity' with which the opinions were recorded, making for 'a most informative and positive contribution to the consultation', it took no

---

[773] Ibid 38.
[774] Ibid 18.
[775] Ibid 41.
[776] A majority asserted that they are reasonably informed about GM crops, while 84.8 per cent were not confident that GM crops are carefully regulated.

apparent effort to identify, understand or articulate the ethical implications of the lack of public trust in risk mechanisms, or in the quasi-exclusive reliance in the scientific community for regulation.[777] It is significant that the Irish Council undertook a public consultation through questionnaires and other methods, and recorded that 'the great majority of those responded clearly displayed a strongly negative attitude towards GM'.[778] However, an avoidance of discussing the ethical implications for GMO regulation of the lack of public trust in these approaches is inadequate. The Council could have ethically reflected and justified these approaches, since otherwise it is merely restating current regulatory approaches without any additional reflection on the very elements of which the public are sceptical. There should have been a systematic reflection and reasoning about the ethics of predominantly relying upon the scientific community itself. This would include the scrutiny of regulatory approaches regarding issues that the scientific community characterizes as side-effects. This is because other social groups that are marginal in regulatory deliberation might not agree, given that the impact adopting GMOs could have on them might be grave.[779]

Three important omissions in the general approach of various GMO-ethics committees in the 1990s are also identifiable in the Irish Council report. These points pertain to:

1.   a lack of attention to the instability of underlying scientific knowledge that is used in regulation of technology,
2.   the need for a deontological ethical scrutiny of risk regulation, and
3.   the necessity to focus on the ethics of research and development of specific technologies.

These points are elaborated below through review of literature that examines important public GMO-ethics reports that pre-date the current regime.

Further, 86.3 per cent agreed that GM crops interfere in nature more unacceptably than prior forms of cultivation, and 78.6 per cent did not trust the governmental and scientific information on safety: the Summary of responses, ibid.

[777]   Ibid 42.
[778]   'Forward', ibid 2.
[779]   See the treatment of the Irish Council on unanticipated side-effects in n. 681. See further, text to nn. 314, 492.

6.3.1   Instability of Underlying Knowledge

Despite systemic issues of scientific ambiguity and ignorance visible in high profile failures like BSE, public concerns about the limits in reliability of scientific knowledge used in regulation are generally left out of ethical scrutiny by most committees. These committees are criticized for their specific approach in disregarding scientific ignorance as an issue of ethical scrutiny. Under this approach, scientific ignorance needs no further ethical elaboration if the scientific community is examining the accompanying hazards during risk assessment. Such an approach is visible in the ideal-typical understanding of public concerns in the advice of the Irish Council. It is also identifiable in the report of the UK Biotechnology and Biosciences Research Council (BBSRC), the UK Nuffield Council advice and the GAEIB report, which are discussed later in this subsection.

Based on the findings of a number of qualitative research projects regarding public perceptions on GMOs,[780] Wynne asserted that the public is concerned about a range of moral concerns beyond the consequences of the use of GMOs.[781] He focused on specific moral concerns like playing God and disrespect of nature, and asserted that the public concerns signified a 'relational dimension' regarding the response of regulators and scientists to the inadequacy of current scientific under-standings. In other words, public concerns are also about the lack of acknowledgement of the limits of scientific knowledge used for the characterization of adverse consequences by the scientific community, in view of prior regulatory failures in not anticipating grave and unforeseen consequences. Through the aforementioned qualitative studies on public perception, he elaborated that while the public is concerned about the quality and lack of stability of the underlying scientific knowledge for GMO regulation, the public is not demanding or expecting a situation free of uncertainty or a freeze on innovation until that moment of clarity. However, it is concerned about the lack of reflection about the quality of

---

[780] See further: Centre for the Study of Environmental Change, *Public Attitudes towards Agricultural Biotechnologies in Europe: Final Report of Project with Five Partner Country Teams (Spain, Italy, Germany, France and the UK)* (EU DG-Research 2000); Sabine Louet, 'EC study reveals an informed public' (2001) 19 *Nature Biotechnology* 15; Robin Grove-White and others, *Uncertain World: Genetically Modified Organisms and Public Attitudes in Britain* (Centre for the Study of Environmental Change 1997).

[781] Brian Wynne, 'Creating public alienation: Expert cultures of risk and ethics on GMOs' (2001) 10 *Science as Culture* 445.

knowledge at issue and, importantly, a failure to relate instability of this knowledge to the purpose of innovation. Wynne stressed that these qualitative studies demonstrate a public expectation that regulators go beyond claims of promoters and check whether the purpose of innovation is sufficiently important to accept inevitable scientific ignorance. That is to say, citizens expect regulators to be permissive during stages of scientific ignorance only if the answer to this question is sound and commensurate with recognized public values and interests through a comparison with alternative available technologies.[782]

Given this public expectation, Wynne found that the resultant public concerns about the ethics of the instability of underlying knowledge cannot be left to scientific communities only, but should also be scrutinized by ethical bodies.[783] For this reason, he emphasized that ethics about technologies like GMOs is not just about consumer concerns that could be solved through individual choice and labelling, but involves wider public concerns whose scrutiny cannot be avoided by invoking individual choice as a framework for solution. Such concerns of citizens about regulators playing God, disrespecting nature or ignoring instability of scientific knowledge were largely understood by earlier GMO-ethics committees as private concerns. In contrast, critics argue that they are publicly articulated assertions and concerns regarding the ethics of ignoring the instability of underlying scientific knowledge by regulators.

While substantiating this point, Wynne is critical of the discussion of disrespect for nature in the UK BBSRC report,[784] 'as one of asking solely about the ethical status of the manipulation', ignoring the 'institutional denial of lack of control, and of dismissing unpredictability'.[785] Since the institutional denial and dismissal of unpredictability itself might be fuelling concerns of disrespect for nature, it needs further ethical scrutiny and elaboration from public ethics committees. He points out how such a treatment in the BBSRC report is possible only because ethical concerns of citizens are deemed to be shorn of intellectual judgement, and by the seamless disregard of the intellectual issue of limits of knowledge.[786]

---

[782]   Ibid 466–467.
[783]   Ibid.
[784]   Biotechnology and Biosciences Research Council, *Ethics, Morality and Crop Biotechnology* (prepared by Roger Straughan and Michael Reiss, UK Department of Trade and Industry 1996).
[785]   Wynne, 'Creating Public Alienation' (n. 781) 461–462.
[786]   See further, ibid 465.

Wynne identified a similar deletion also in the UK Nuffield Council report,[787] which he strongly criticized:

> [I]t ignores the very issue raised by typical public concerns – whether we should assume that science indeed reliably identify future consequences, or whether to the contrary, there are going to be consequences of which current knowledge is ignorant, or which are contingent on so many independent conditions that we can only say they are conceivable but with unknown likelihood.[788]

From Wynne's elaborations it is clear that the regulatory treatment itself is understood as the object of public concerns in playing God and disrespect for nature. Then, the object of public scrutiny is a question of institutional ethics about the relational aspect between the public purpose of innovation and the instability of scientific knowledge. Understanding them as private concerns is denying a profound public influence on shaping public values in regulation of GMOs.

This pattern of ignoring ethical concerns about scientific incertitudes through reframing them as only risk assessment, and referring them back to the scientific community, recurs among earlier GMO-ethics committees. It is also discernible in the advice of GAEIB to the Commission in 1995, which identified the primary ethical imperative applying to all foodstuffs as being related to their safety.[789] It stated that 'when such products are authorized to be placed on the market *they will have already met* the required safety standards both for human health and for the environment; *these standards*, which are also subject to ethical considerations, *do not fall within the scope* of this opinion'.[790] It is through this deletion of an important question of institutional ethics that the report focuses on the benefits of biotechnological progress as 'a new element that can contribute towards meeting food requirements in Europe and throughout the world'.[791]

In these illustrations, the seamless writing out of public concerns about the lack of reflection regarding intricate connections between limits of underlying scientific knowledge and judgements regarding social value

---

[787] Nuffield Council for Bioethics, *Genetically Modified Crops: The Ethical and Social Issues* (Nuffield Council 1999) <http://www.nuffieldbioethics.org/sites/default/files/GM%20crops%20%20full%20report.pdf> last accessed 24 September 2011.

[788] Wynne, 'Creating Public Alienation' (n. 781) 465.

[789] GAEIB (n. 687).

[790] Ibid 3. Emphasis added.

[791] Ibid.

and utility in permitting new innovations is apparent. It is through such deletion that ethical issues about playing God and tampering with nature are reduced as emaciated personal concerns of consumption, which can then be resolved through labelling measures under the logic of consumer choice.

6.3.2 Ethical Scrutiny of Risk Regulation

Quite apart from the reliability of the data in risk analysis, there is a lacuna in the approach of various public bioethics committees in failing to scrutinize the ethical aspects behind the framing of the regulatory tool of risk. Various bioethics committees have taken the normative assumptions that are implicit in technical risk analysis as natural, and not in need of ethical scrutiny. This failure by public bioethics committees like the Irish Council to scrutinize the normative assumptions in risk analysis ignores societal disagreements about them, and thus implicitly denies the representational inadequacies.

The need for public bioethics committees to scrutinize and discuss the normative assumption in risk analysis is emphasized by Levidow and Carr. They demonstrate the incompatibility of the various accounts of the risk problem in the GM debate, where each set links risk and opportunity to its own account of benefit, an account that necessitates further ethical scrutiny:

> [F]or some critics, the risk problem goes beyond physical measurable effects (*consequences/cost*), to encompass features that some proponents regard as benefits. Such features include the tendency to treat society's problems as genetic deficiencies of organisms, and to treat nature as a commodity ... for example, many people perceived biotechnology as a profit-driven force that threatens their ethical-cultural values.[792]

Through research interviews, the authors assert that when regulators set priorities for risk research, judge available evidence and/or seek additional evidence for the treatment of some potential effects as plausible and unacceptable, this also involves considering the ultimate benefits of a particular release: 'if doubtful about benefits, then they may set more stringent criteria'.[793] 'From one stance, society is at risk from failing to reap the indispensable benefits of biotechnology. From other stances,

---

[792] Les Levidow and Susan Carr, 'How biotechnology regulation sets a risk/ethics boundary' (1997) 14 *Agriculture and Human Values* 29, 33.

[793] Ibid 37.

society is at risk from biotechnology, which precludes beneficial alternatives'.[794] The authors elaborate how competing environmental values
entered most aspects of risk assessment in the 1990s: while in the UK,
'harm was meant as that which infringes farmers' property rights or
impairs the agro-chemical control of weeds', in some European countries
there was a broader definition of environmental harm as 'any effect that
could preclude future options for sustainable agriculture, including ...
ecological uncertainty from unacceptable effects of large-scale commercial use'.[795]

If the assumptions that go behind risk frames in risk analysis and risk
management are implicitly value laden, then such assumptions need
public scrutiny and ethical elaboration. The approach in pieces of advice
like that offered by the Irish Council is conspicuously inadequate in this
regard.

6.3.3   Ethical Scrutiny of Technological Trajectories

The third major lacuna identified in the traditional scrutiny by public
GMO-ethics committees is their lack of attention to the ethics of research
and development of GMOs. Public ethics may need to scrutinize not only
the ethical concerns regarding the use of GMOs, but also the ethics of the
directions of biotechnological research. This is because the public may be
concerned about a range of moral concerns and not just the consequences
of the consumption of GMOs, as evidenced in the studies mentioned in
the previous sections.[796]

An implicit acceptance of the ethical judgements of developers of
controversial technologies is in opposition to the need for an open ended
and explicit public discussion about the direction of such research and
development, including through bioethical scrutiny of public bodies. This
is since 'in reality, the research itself is already shaped by values which
link the production of scientific knowledge...to notions of progress and
strategic interest'.[797] Once these values are taken for granted and beyond
the scrutiny of public bioethics, Carr and Levidow argue, expert committees merely mitigate the undesirable effects of the prior value-choices,

---

[794]   Ibid 36.
[795]   Ibid 38.
[796]   See n. 780, text to nn. 774–776.
[797]   Levidow and Carr (n. 792) 31.

which are already embedded in techno-scientific research and development programmes.[798] In the absence of this scrutiny, GMO-ethics is seen only to conduct a risk–benefit analysis by restricting public concerns as individual moral feelings. This is in opposition of a public institutional ethics where the production and application of biotechnological knowledge, as well as the implicit ethics involved in the regulator's judgements, are explicitly scrutinized by public bioethics committees.[799]

In any case, an explicit recognition of normative and ethical commitments that shape the development of technologies is currently absent during risk regulation. This makes an examination by public bioethics committees of the ethics of the normative commitments that fuel specific trajectories of technological development even more imperative. This is connected to the demand for a public debate about the benefits from GMOs, which is part of the public contestations about the technology to influence the regulation of GMOs. In the absence of an explicit public scrutiny about benefits regulators may assume that 'any product brought

---

[798] Ibid. See further: Michael Dillon, 'Sovereignty and governmentality: From the problematics of the "New World Order" to the ethical problematic of the World Order' (1995) 20 *Alternatives* 323. He described ways in which a discourse of addressing 'ethical concerns' by expert committees have served to defend technocratic commitments silently embedded in the larger policy culture, and simultaneously renders those commitments invisible. Further, Wynne pointed out another area where the technique of asking the question 'where to draw the line?' has been employed – 'for instance, since all science is reductionist, or history has demonstrated how all new technologies inevitably have far-reaching effects, crop biotechnology cannot then be singled out as the sole target for moral censure on these grounds'. He reminded the reader that, since in practice far reaching effects mean lack of intellectual control by science over future consequences, it is inappropriate that the ethical consequences of such development of technologies are not publicly scrutinized by the bioethics committees. Wynne, 'Creating Public Alienation' (n. 781) 462.

[799] Carr and Levidow highlighted a report commissioned by the European Parliament that identified wide-ranging value judgements as ethical issues that need scrutiny, including instrumental transformation of nature, inequitable distribution of risks and benefits, the significance of ecological uncertainty, and the cognitive frameworks involved in risk assessment. Using the term 'bioethics' to include the implicit policies of R&D and regulatory priorities, in evaluating the potential benefits, the report contrasted the commercial aims and practices with the humanitarian claims of the development of agri-biotechnology, through which there was an attempt at a fundamental ethical scrutiny. Levidow and Carr (n. 792) 38.

forward by any promoter for regulatory decisions, by definition equals prospective social benefit because that free entrepreneur has defined it as such'.[800]

Taking stock of the ramifications regarding the representational issues: first, given the lack of trust in the scientific establishment there is a need to recognize the limitations in the assumption that risk analysis is the only conceivable way to answer public concerns about the instability of underlying knowledge. This in turn calls for further ethical scrutiny and elaboration of how to deal with the instability of scientific knowledge during regulation. Secondly, symptomatic of the criticism about the advice of the Irish Council, the normative assumptions during the framing of risk need to be publicly scrutinized by bioethics committees, rather than assuming those implicit normative judgements to be natural, unproblematic and unnecessary of ethical elaboration. Devoid of these two factors, GMO-ethics could be imagined and employed in a way that effectively keeps risk and ethics as two separate domains. This could effectuate the avoidance of public scrutiny of the normative assumptions of risk regulators, as well as the instability of scientific knowledge used for risk regulation. Finally, implicit judgements during the development of specific trajectories of technologies, including GMOs, about social benefits, progress through innovation and the conceptualizations of the good life need explicit scrutiny by public bioethics committees. These include a public scrutiny of the reasons through which specific techno-scientific developments are found as desirable, explicit scrutiny of the social values they pursue and the ways in which such developments serve the public good. Thus the suggested appraisal should be in comparison with alternative trajectories and relate judgements of social value with possible dangers, keeping in mind the fundamental instability of under-lying scientific knowledge. With these criticisms in the foreground, the next section examines the advice of the EGE and the Danish Council, including the positive manner in which these criticisms appear to be taken on board.

---

[800]   Brian Wynne, 'Elephants in the room where publics encounter "science"? A response to Darrin Durant' (2008) 17 *Public Understanding of Science* 21, fn 16.

6.4 ETHICAL SCRUTINY OF PUBLIC CONCERNS: ADVICE OF THE EGE AND DANISH COUNCIL

Pursuant to a request by EU President Barroso, the EGE prepared an advice to address new challenges and opportunities for EU agriculture accompanying modern developments in agriculture technologies.[801] The scope of the opinion was broader than GMOs and included ethical issues concerning food security, safety and sustainability regarding the primary production of food of plant origin. However, EU agricultural policy areas like fisheries, livestock farming and green biotechnology for pharmaceutical use were explicitly excluded. The consultations that preceded the finalization of this opinion involved a range of actors including individual experts, representatives of European and international institutions, and representatives of civil society. It culminated in a round table comprising invited speakers, EGE members and around a hundred invited delegates from different interest groups, including the agro-research industry, consumer organizations, religions, environment organizations, food and farm industries, and representatives of EU Member State ethics committees.[802]

6.4.1 The Need for Technology Impact Assessments

The EGE report recommended the pursuit of an integrated approach in agriculture involving a continuous assessment of the balance between expected outcomes and required inputs (like resources and energy) at the technical level. It identified human dignity and justice as two fundamental ethical principles for the EU, and called for explicit embedding of them in EU agricultural policy, including in the production and distribution of food products.[803] It found a scenario where members of society interact and act on the basis of commonly held values as desirable, and

---

[801] European Group on Ethics in Science and New Technologies, *Opinion No. 24 on Ethics of Modern Developments in Agricultural Technologies* (EGE 2008) <http://ec.europa.eu/bepa/european-group-ethics/docs/opinion24_en.pdf> last accessed 12 November 2011.

[802] European Group on Ethics in Science and New Technologies, *EGE Proceedings of the Roundtable on Ethics of Modern Developments in Agricultural Technologies, 18 June 2008* (EGE 2008) <http://ec.europa.eu/bepa/european-group-ethics/docs/publications/agriculture_technologies_ethics.pdf> last 12 November 2011.

[803] Ibid 48.

advocated a continuous assessment to adhere to its ethical goals.[804] The approach recommended by the EGE included the recognition of the importance of principles of food safety and the right of access to food for all, as well as the need to implement a different model of agriculture that is sustainable and multi-functional. It argued for a model that includes the use of various principles like stewardship of the land, preservation of the resource base, the health of farm workers and preservation of small biota rich in biodiversity, as well as the values of rural communities and agricultural landscapes.[805]

Importantly, in addition to risk assessment, the report emphasized the need for impact assessment of technology, at both national and European levels.[806] It recommended that the effects of a technology should be carefully studied and evaluated by means of an impact assessment that involves a comparison of existing and new technologies, before it is considered for use in agriculture.[807] This important recommendation about impact assessment, which has a significant bearing on the regulation of GMOs, needs to be read along with the committee's recommendation on public participation. The report recognized that agriculture as a policy domain ought to be inordinately consistent with societal needs, goods and expectations, given its significant role in terms of economics, labour and social goods.[808] Therefore the advice emphasized the imperative role of public participation and effective involvement of all stakeholders.[809] Further, it sought an integrative action plan that emphasizes the importance for agricultural regulation to address the needs of local and regional markets. It recommended that the plan should include local and regional transport systems, health and education infrastructure, and systems of accountability of political institutions and big corporations.

Within this broad framework, the report made specific recommendations about a number of issues, including the use of GM crops, agricultural biodiversity, sustainability of agricultural technologies and technology impact assessment of agricultural technologies.[810] Significantly, the report understood the relevant ethical concerns as ones that cannot be confined only to policy design, but also deeply concern the technological dimension of modern agriculture. This included unexpected

---

[804] EGE, 'Opinion 24' (n. 801) 62–63.
[805] Ibid 63.
[806] Ibid 61.
[807] Ibid 62.
[808] Ibid 66.
[809] Ibid 67.
[810] Ibid 60–66.

consequences that may arise out of the use of new technologies in agriculture due to the rapid and fundamental impacts of these technologies.

Underlining the need to strike a balance that ensures food security, decent and dignified employment, and health and respect for the environment as key,[811] the report linked the question of promotion of new technologies in agriculture not only to its outcomes, but also to the modes of ascertaining those outcomes, including questions of 'for whom, and for how long'.[812] The report outlined the various axes of intense polarization around the cultivation of GMOs, including the differences in regulatory approaches in GMOs between the EU and the US or Argentina, as well as the sharp divisions within the EU, say, between industrial stakeholders and consumer organizations or environmental protection organizations. Within this rubric, it emphasized the demand from many quarters for an environmental impact assessment of new technologies, and a technology impact assessment regarding the dangers of implementing them, compared with persisting with existing agricultural technologies.[813] After recognizing the existing obligations for a scientific risk assessment, the report emphasized the importance of the precautionary principle and the investigations regarding unwanted pleiotropic effects (related to the effect of a single gene having multiple effects in different tissue formations).[814] Importantly, it also recommended that risk management procedures should be revised to take full account of the need for an impact assessment of all new technologies.[815]

6.4.2 Protection of Farmers' Right to Save Seeds

Recognizing principles of fairness, justice and solidarity as fundamental in the regulation of world trade, the EGE raised concerns about the accentuation of market dominance within a few companies over much of agricultural production, through the existing Intellectual Property Rights (IPR) regime for plant varieties and GM crops. The report was concerned about the aforementioned disregard of the accentuation of market dominance within a few companies over much agricultural production, in current IPR policies. It cited the increasing monopolization of the food, seed and agro-chemical market by a small number of companies. It called

---

[811] Ibid 58.
[812] Ibid 52.
[813] Ibid 56.
[814] Ibid 64.
[815] Ibid.

for a proper balance between WTO rules and the socio-economic aspects of different regions of the world, a balance which is guided by concerns of fair trade, justice and solidarity in global trade in agricultural products, including in seeds.[816]

Here, the report emphasized the need to maintain farmers' rights to save seeds and use them in subsequent seasons. It drew attention to grave concerns arising from the current industry practice of standard licence agreements (especially) in the sale of GM seeds, through which seed companies controlled the use of saved seeds.[817] Noting that concerns of economic monopoly and biosafety have been raised in the context of the use of GMOs, it underlined that only those agricultural technologies that are conducive to the goals of the advice, including the ethical criteria indicated in it, should be sustained in the EU.[818] While recognizing the need to promote innovation in agriculture, it called for specific efforts to mainly support technologies that are conducive to food security, safety and sustainability. This was in order to ensure ecologically and socially sound agricultural production (techniques and methods) based on fair treatment both of the environment and of farmers.[819] It further encouraged the EU to increase the budget for research in agricultural sciences, green biotechnologies and all other sustainability oriented agriculture research, to achieve the ethical goals articulated in the opinion.[820]

The EGE report is conspicuous by its broad focus of ethical scrutiny regarding the regulation of GMOs and other agricultural technologies. Such an emphasis is also manifested in the focus on the issue of growing corporate control over the farm sector, particularly through the employment of standard licence clauses to stop the traditional practice of using saved seeds by farming communities. It thus moves away from relying on a narrow scientific frame of risk of earlier GMO-ethics, elaborated in the previous section. Even the emphasis on the precautionary principle and the stated need for investigating unwanted pleiotropic effects is within this broad approach, since it connects social value and impact of the technology to risk regulation. This broad ethical frame on GMO regulation seeks to address the representational question by taking public contestations seriously, and by imagining space for wider public participation. This is visible in its insistence on technology impact assessments, and the imperative role of public participation in technology impact

---

[816]   Ibid 59.
[817]   Ibid 68.
[818]   Ibid 64.
[819]   Ibid 62.
[820]   Ibid 66.

assessments. Such a broad approach is also visible in the advice of the Danish Council of Ethics, which is described next.

6.4.3 Advice of the Danish Council of Ethics

The Danish Council of Ethics prepared a report in response to a request from the Danish Environment Minister to identify 'the more intangible subjects that have a significant bearing on the public debate on the use of genetic engineering in the field of food, beyond risk evaluations'.[821] The report reviewed the debate about the spread of GM plants, the legislation linked to the regulation of GM crops, and the appropriate regulatory responses.

The report underlined the importance of examining both direct and indirect impacts of GMOs, including implications for the environment and health of non-target organisms, and the farming structure and economy. It recognized that an evaluation of the utility and hazard of a particular plant has to be preceded by identifying the values relevant to such examinations. It emphasized the difficulties in making such a regulatory call, while representing various points of view to the fullest possible extent.[822] Further, it examined the possibilities of incorporating ethical considerations in public decision-making, through various vantage points of utility assessment, religious views, eco-centric views and 'decision theories'.[823] Identifying utility as a central issue, it underlined the importance of conducting utility assessments through a number of considerations. These considerations included sustainability and consideration for posterity, distribution of goods, agricultural developments and risk evaluation. The report insisted that this assessment has to be coupled with answering the question as to who receives the maximum benefit from such technological improvements.[824]

The report noted that scientific risk evaluation currently determined the decision to release, and emphasized the need to include utility in the evaluation during the decision to release. By citing various European studies about citizens' attitudes to GMOs, it underlined that citizens' perceptions that GMOs benefit them play an essential part in consumer

---

[821] Danish Council (n. 688).

[822] Ibid 37.

[823] Ibid 73–96. The 'decision theories' (theories concerned with identifying the issues relevant in a given context, including its values and uncertainties) that the report examined are the precautionary principle and cost–benefit analysis (ibid 89).

[824] Ibid 62–65.

approval. This is since scepticism is absent in areas like GM medicine, disaster management and decontamination, or famine reduction. It also underlined that citizens object to large multinational corporations reaping inordinate benefits from large scale use of GM seeds, despite persisting uncertainties.[825]

The report examined the concept of utility, its incorporation in regulatory decisions and the problems that arise due to a wide spectrum of interpretations of utility.[826] While it felt that utility assessment is desirable, it found the discussion tricky due to lack of agreement about objective standards regarding usefulness, as well as the uncertainty surrounding both beneficial and undesirable effects from the use of GM crops. The Council found that the difficulty in utility assessment is not merely about predicting the direct and indirect risks and the useful effects brought about by GM cultivation, but further, the consequences of the actual application depend upon a number of social and human factors that are difficult to assimilate into the theoretical prediction for consideration of utility.[827]

6.4.4   Characterization of Dissent about Utility

The report emphasized the need for a holistic evaluation of GM crops, and approached its recommendations regarding considerations of utility in approval procedures for release in a salient manner. The advice characterized three different points of view among the committee members regarding a determinative linkage between permissions to release a GM crop and its utility assessment.[828] A minority of members felt that considerations of utility should not be directly included in the regulator's approval procedure. This was since the evaluation of any particular GM crop's utility will invariably be so elastic and subjective by nature that it is best left to the market, and the demands of the consumer. In this viewpoint, the consumers' view of utility ought to determine the production and distribution of particular GM plants, provided that it entails no risk to humans and the environment.

---

[825]   Ibid 55.

[826]   The report examined the liberal model, the utility-based and communitarian views here, ibid 97–105. It explored different models for correlation between GM plants, utility, ethics and belief, and also discussed the possibility of using 'satisfaction of basic human needs' as an indicator of utility.

[827]   Ibid 110.

[828]   Ibid 112–115.

In contrast, a big majority argued to include considerations of utility in the approval procedure for release, although members within this group disagreed about the precise role of utility in the procedure. Bearing in mind the great effectiveness of GM methods, since it allows quick and highly extensive modifications, all members of this group were concerned about inadvertent negative consequences and shifts of ecological balances. Some members within this group felt that GM crops can only be approved for release if it can be demonstrated with great probability that only minimal risk is associated with their use. Along with this criterion, these members also insisted that it must be shown with reasonable probability that their release may entail substantial benefit or essential utility, 'i.e. must either be able to satisfy fundamental human needs, or remedy essential environmental problems, where this is not possible in any other way'.[829] Citing secular and religious rationales, this group held that the integrity of nature should be respected as a basic premise. Consequently, one should be cautious about altering existing species or creating entirely new species regardless of justifications provided, such as the interests of mankind. Others within the majority felt that it is possible to weigh up utility and risk, and that GM crops can be approved if the risk of using them is deemed to be entirely minimal, even when no additional utility is anticipated. They nevertheless felt that discretion should be exercised in modifying existing plants, given the limitations of foreseeing the consequences of using them.

6.4.5   Restrictions on Farmers' Right to Save the Seed

Regardless of the divergent views mentioned above on how utility and risk assessments should be weighed, a large majority of members underlined an important economic and social factor regarding the implication of continuation of the use of farm-saved seeds. This is because currently 'once the farmer has bought a portion of GM sowing seed, according to customary practice he is not allowed to use his crop as farm-saved seed without paying a fresh fee to the patentee'.[830] The report pointed out that this aspect makes a vital change in farming culture. The Council unanimously emphasized that this practice raises grave implications for small farmers and farming culture, and underlined the need to include this factor in the debate on GM crops.

---

[829]   Ibid 112.
[830]   Ibid 116.

It is significant that the disagreement about employment of utility within the members of the Council, a central aspect of public values about the use of GM technology, is scrupulously and remarkably recorded, along with all the salient viewpoints and their respective rationales. There was an absence of ethical scrutiny regarding the development of food technologies itself in the report. Nevertheless, the characterization of conflicting rationales, regarding utility as a central value independent of scientific risk, is an appropriate exercise of the public function by the bioethics council. Through this, the advice has fostered public debate, identified rationales of ethical convergence and divergences in society, and ethically scrutinized the regulatory positions regarding divergences.

Both the EGE and the Danish Council acknowledged the grave concern about the control of the traditional practice of saved seed, through licence agreements by GM seed companies, and sought further public debate about this in the GM debate. They also identified increasing concentration of control over agricultural production by a small group of for-profit corporate entities, and identified public concerns that connected appraisals of profits, benefits and utility to regulatory investigations in risk; thus seeking to include assessments of utility and impact of the technology in regulation of safety concerns. The EGE sought technology impact assessments, and environmental impact assessment, over and above the existing requirement for risk assessment prior to release of GMOs. It further put focus on ethical concerns regarding the development of new technologies in relation to the lack of stability of the underlying scientific knowledge. Both committees recommended that risk management measures should take full account of the need for impact assessment. This holistic approach is in contrast to the restricted approach of the Irish Council and earlier traditional GMO-ethics. Though there was no lay representation granted by law in all the three committees, they involved public consultations before the preparation of the reports. The Danish advice was conspicuous by its representation of a range of minority opinions. The identification of impact and utility assessment in the EGE and Danish reports also opens avenues for bringing public concerns of social value, and contestable notions of justice and welfare, to the centre of the regulatory debate, and further recognizes public debate and participation in its conceptual core.

6.5 TOWARDS PUBLIC DELIBERATION OF PUBLIC VALUES IN GMOS

Before we conclude this chapter, this section focuses on the possibilities opened up for public participation in EU regulation of GMOs through the recommendations for impact assessment in the EGE and Danish reports. These two reports provided a broader range of scrutiny in comparison with the scrutiny by earlier public GMO-ethics committees, which restricted their gaze within a strict scientific conceptualization. This section highlights the recommendations of these reports for environmental impact assessment, utility assessment and technology impact assessment of the development and use of GMOs to see if they can provide for public participation in EU GMO regulation. The intention of the section is not to offer them as policy recommendations to make EU regulation more participatory, but merely to evaluate the room for participation that such assessments offer if such impact assessments were to be implemented in the EU regime.

Technology assessment is understood as an analysis of all significant primary, secondary, indirect and delayed consequences or impacts present and foreseen of a technological innovation on society, the environment or the economy.[831] Within this definition, it is evident that risk assessments, environmental impact assessments and utility assessments fall well within the ambit of this exercise. Further, the proposal to add a safeguard clause for Member States' decisions on cultivation of EU-authorized GMOs (under the Directive) requires 'social, economic and ethical aspects' to be taken on board.[832] This has the potential to be a form of technology assessment. There is a cognitive superiority in invoking technology assessment as a regulatory practice over an emphasis on safety through risk assessment. This is since it can be a more holistic tool to connect all potential scientific, social and environmental outcomes regarding the development and use of a new technology. However, whether technology assessment can be a channel of representation of public values in which

---

[831] US House Committee on Science and Astronautics, 'Office of Technology Assessment background and status', Committee print, August 1973, cited in Walter Hahn, 'Technological assessment and the Congress' [1975] January/February *Public Administration Review* 76, 77. See further: Ron Westrum, *Technologies and Society: The Shaping of People and Things* (Wadsworth Publishing 1991) 325, where technology assessment is defined as an attempt to predict what the effects of technology will be if it is implemented.

[832] Commission Communication (n. 141).

groups broader than expert communities can participate and be represented is unclear. This depends on the institutional and conceptual imaginations within which technology assessment is implemented as a regulatory tool.

Currently there is no explicit requirement for technology assessment in the EU regulation of development and use of GMOs, let alone legal specifications for public participation in such exercises. Technology assessment first gained prominence in the 1970s, through the creation of the Office of Technology Assessment (OTA) for providing objective analysis to the US Congress.[833] The OTA was expected to 'provide early indications of the probable beneficial and adverse impacts of the applications of technology' to assist the US Congress.[834] The experience of the OTA (until it was wound up in the mid-1990s)[835] is traced by some commentators as a significant influence in many EU Member States.[836] There is a plethora of bodies that provide such advice for parliamentary bodies in Europe, including the Rathenau Instituut in the Netherlands, the Office of Technology Assessment at the German Parliament, the Catalan Foundation for Research and Innovation for the Parliament of Catalonia, the Danish Board of Technology, and the Institute of Technology Assessment of the Austrian Academy of Sciences. The Scientific Technology Options Assessment panel has organized the process of technology assessment consultations in the European Parliament since the late 1980s. This European parliamentary panel has been advised by the European Technology Assessment Group since 2005, which is a network of earlier-mentioned institutions that provide technology assessment advice to Parliaments in the Member States. The Commission has funded various research projects in this area, including FAST (Forecasting and Assessing for Science and Technology), Formakin (Foresight as a Tool for the Management of Knowledge Flows and Innovation), Europta (European Participatory Technology Assessment), ASTPP (Advanced Science and Technology Policy Planning), ITSAFE (Integrating Technological and

---

[833]   2 U.S.C. § 471 (a).

[834]   2 U.S.C. § 472 (c) (2006).

[835]   The OTA was wound up by the Congress as part of the Republican cutbacks in 1994. See for more on the OTA: Bruce Bimber, *The Politics of Expertise in Congress: The Rise and Fall of the Office of Technology Assessment* (State University of New York Press 1996).

[836]   Tellichini (n. 744) 285–287, fns 3–5.

Social Aspects of Foresight in Europe) and TAMI (Technology Assessment in Europe; Between Method and Impact).[837] How these EU research projects may affect the regulatory reality on GMOs is currently indeterminable.

There are calls for bringing explicit technology assessment into the centre stage of regulation of new technology with an explicit emphasis on public participation in such assessment of technology.[838] The critique of expert-dominated technology assessments runs parallel to the calls for public participation in risk assessment, conceptually an integral part of assessing the impact of any new technology.[839] This emphasis on public participation is explicit in the EGE recommendation for impact assessment of new agricultural technologies like GMOs. The role of the Danish Board of Technology, an avid advocate of pTA (participatory Technology Assessment),[840] and the general culture of consensus-seeking through citizen-based deliberative exercises in the Danish political system are significant in making sense of the recommendations for utility evaluation

---

[837] See further: the Commission archives in <http://cordis.europa.eu/foresight/research.htm> last accessed 5 January 2011.

[838] See for instance, Editorial, 'Open to all' (2010) 465 *Nature* 10; Daniel Sarewitz, 'World view: Not by experts alone' (2010) 466 *Nature* 688. See further: Michael Rodemeyer, Daniel Sarewitz and James Wilsdon, *The Future of Technology Assessment* (Foresight and Governance Project, Woodrow Wilson International Center for Scholars December 2005) <http://www.wilsoncenter.org/sites/default/files/ techassessment.pdf> last accessed 5 January 2012; Richard Sclove, *Reinventing Technology Assessment: A 21st Century Model* (Science and Technology Innovation Program, Woodrow Wilson International Center for Scholars April 2010) <http://www.wilsoncenter.org/sites/default/files/Reinventing TechnologyAssessm ent1.pdf> last accessed 5 January 2011; <http://www.loka.org/TrackingConsensus.html> last accessed 5 January 2011.

[839] When 'the ends of knowledge are power, science becomes the servant of the powerful': Hazel Henderson, 'Philosophical conflict: Reexamining the goals of knowledge' [1975] January/February *Public Administration Review* 77, 79; it was suggested that prudence demands that all technology assessment 'ought to include an inventory of vested interests in the promotion or suppression of any technology to be assessed and an analysis prepared of the force field that such vested interest may create': Francois Hetman, *Society and the Assessment of Technology* (OECD 1973) 85. Public participation is central to making such inventories. See further: Sherry Arnstein, 'A working model for public participation' [1975] January/February *Public Administration Review* 70, 71.

[840] See 'The Danish Board of Technology wins international award for public participation' (15 March 2011) <http://www.tekno.dk/subpage.php3?article=1735&survey=15&language=uk> last accessed 21 January 2011.

by the Danish Council.[841] A question of how to involve the lay public in an institutional sense cannot be addressed here since matters of institutional design are outside the purview of this book. Having noted the promise of technology assessment, including environmental impact assessments and utility assessments, for its broader focus (in comparison with a techno-scientific conceptualization of risk assessment), this section has reiterated that whether such assessments can be incorporated as regulatory tools in EU GMO regulation is currently indeterminable. This is since it depends on the legal and institutional context through which technology assessment as a regulatory tool will be implemented, a context which is currently speculative.

Taking stock, the focus of this chapter pertained to EU regulatory strategies regarding pursuit of public values during the regulation of the development and use of GMOs, and the role of public participation in them. The strategy of consumer autonomy and ethical pluralism employed through labelling and coexistence rules can sufficiently involve consuming publics only if the disagreements about values pertain exclusively to the employment of a technology, and not to its development. Further, an attempt at creating an ethically neutral framework to pursue ethical pluralism itself is beset with ethical controversies. These include the question as to which set of public values requires statutory ethics labelling, as well as enumerated technical and conceptual controversies about coexistence strategies. Controversies about coexistence mechanisms include doubts about the possibilities of an effective segregation for different agricultural forms to coexist. This generates significant doubts among some sections that a stated attempt at creating a neutral framework is in fact the road toward fait accompli situations in which organic and conventional agriculture will have to live with the presence of a certain amount of GM. Such a framework will not be seen as ethically neutral by these groups, exposing serious limitations of labelling as a predominant strategy to address disagreements regarding public values about regulation of GMOs.

After identifying the second regulatory strategy of advice from public bioethics committees, the chapter emphasized the necessary connections of public bioethics to the public sphere, as opposed to restricting itself to methodological justifications of analytical philosophy. It noted that the challenge for public bioethics committees is to identify and characterize the full range of public concerns about the ethics of regulation, including the use and development of GMOs as a technology trajectory. In the

---

[841]   Horst and Irwin (n. 467).

examination of relevant reports from three public bioethics committees, major criticisms about the focus and substance of various reports in the previous regime were found to be symptomatic in the Irish advice. These included a singular emphasis on the ethics of consumption of GMOs (as opposed to a scrutiny of the ethics of research and development), a lack of scrutiny of the instability of the underlying knowledge used in regulation, and the avoidance of a scrutiny of the normative questions implicit in risk analysis. Dissent and disagreement regarding public values about development and use of technology among various sections of society need to be made visible within public bioethics committees, particularly since Europe aspires to be a polity that 'represents the values of all Europeans'.[842] Tellichini identified the Commission's public assertion of ethics as an official narrative that amounts to an essential frame for the idea of European citizenship – as a way to represent citizens' values and a glue to move from an economic union towards a political union.[843] She presented evidence from the archives of a Commission web-page regarding the role of GAEIB:

> in setting it up the European Commission has highlighted its desire to integrate Europe's science and technology in a manner that serves the interests of European society and respects the fundamental rights of every European citizen ... European integration must mean more than establishing a single market; progress in science and technology must be given a human, social and ethical dimension, otherwise European citizenship cannot be established.[844]

For the effectuation of such a claim of European citizenship in science and technology, public engagement and wider deliberation have to dominate the expert committee enterprise. Otherwise it would be a process where 'citizens enter this ethical domain more as intended objects than as subjects and actors of ethical decisions, notwithstanding the assumption that ethics represents citizens'.[845]

Further, the supposed ethical turn in regulation seeks the legitimation of the resolution of techno-scientific conflicts increasingly in terms of ethical concepts and categories. While this use of bioethics in regulation is criticized for its continuities with the earlier expert-dominated governance despite its fundamental concern with democratic legitimation, Irwin

---

[842] Expert Group on Science and Governance, 'Taking Knowledge Society Seriously', (n. 53) 52.
[843] Tellichini (n. 744) 295.
[844] Ibid fns 7–9.
[845] Ibid 295.

identified a partial displacement of the technical with the emergence of the ethical.[846] This is a movement from a technical frame developed by deliberation within the scientific community (where technical problems and dangers to health or safety could be discussed, but not questions of social order, power or notions of common good) to a potentially wider mechanism for social appraisal of science.[847] A nuanced defence of the ethical turn itself is possible only within a conceptual framework of public deliberation, as opposed to an expert-dominated committee procedure that, despite taking into account wider public concerns, may remain within a rubric of promotion of technological endeavours by market enterprises. This would suggest that only a full blown deliberation may be able to sufficiently legitimate techno-scientific regulation during public contestations about technology and its specific trajectories.

However, effectuation of such deliberations and public participation within bioethics frameworks needs new kinds of regulatory innovations, including conceptual and institutional arenas where such public deliberations can occur. Significantly, pieces of advice of public bioethics committees under the current regime, viz., the EGE and the Danish Bioethics Council, have brought these issues into focus. Such focus, by itself, does not correspond to the fundamental problem posed by the expert-dominated committees of a better engagement with contestations about GMOs and representation of divergent public values. This is since use of ethics as a separate component in regulation of techno-science itself anticipates the inadequacies of a reliance on the values of techno-scientific communities to make their research and development compatible with a range of public values. The recommendations for technology assessment, environmental impact assessment and utility assessment by the EGE and the Danish Council, respectively, are important openings for a more open engagement with public values about research, development and use of new technologies like GMOs. This is since they do not restrict themselves to a regulatory emphasis on techno-scientific risk as the dominant voice for development and use of GMOs. However, whether such assessments will lead to better representation, either through deeper public participation or in other ways, in the EU regulation of the research, development and use of GMOs, is currently speculative, as these recommendations are not implemented in EU law.

---

[846] Alan Irwin, 'The politics of talk: Coming to terms with the New Scientific Governance' (2006) 63 *Social Studies of Science* 299.

[847] Moore (n. 732) 201.

7. Conclusion

> ... we are always in the position of beginning again.[848]

This book sought to identify the space for public participation in the EU regulation of GMOs, given the problem of representation in liberal democracies regarding governance of emerging technologies. During the discussion in various chapters, the scope for public evaluations of social value and utility of new technologies emerged as an important theme in the regulation of new technologies. Social value was noted to be a crucial element in public appraisals of emerging technologies like GMOs, especially in the face of complex and unstable underlying scientific knowledge that informed risk regulation. Public bioethics committees like the EGE and the Danish Council have emphasized the important connections between the social purpose of a new technology and the public acceptance of basic scientific incertitudes about the development and employment of GMOs. These reports found questions like who would substantially benefit from the employment of such technologies, and whether there are equally viable alternative technologies that involve less scientific instability, as important considerations in the regulation of GMOs. This has important implications for the room for public participation in EU regulation of GMOs.

Risk regulation was found to be the predominant route through which GMOs are currently regulated in the EU. Participatory discourse is often invoked to make the political management of risk more acceptable to the general public. In this book, risk is contended to include public trust and engagement in its conceptual core, in contrast to a bifurcated model of risk that conceptualizes techno-scientific communities as the sole custodian of risk assessment. It was noted that issues of public trust in the techno-scientific characterization of risk do not necessarily pertain to the normativity of science, but regard the norms that underlie the identification of the hazard and the ethics of statistical projections through which the occurrence of hazards is reduced to probabilities. Here

---

[848] Michel Foucault, 'What is Enlightenment?' in James Faubion (ed.) *Essential Works of Foucault 1954–84* (The New Press 1997) 303, 317.

scientific/technical processes are to be honed towards identifying and investigating all potential hazards that are conceivable, including some that are not considered plausible by techno-scientific communities but are found to be of concern within other publics. Such a participatory investigation of incertitudes may constitute an active pursuit of social trust in risk regulation. In contrast, the current framework bifurcates risk regulation into technical assessment and political management in a way that allows little room for the involvement of wider publics. The framing of risk is conducted in a regulatory milieu that is devoid of any formal requirement for techno-scientific communities to involve concerned social groups. This is aggravated by a framework where the data about potential hazards are generated, collated, located and analysed by the private applicants interested in their market use, right from the start of the development of this technology. Be it under the Contained Use Directive or under the Deliberate Release Directive, it is only at the very last stage of formal risk analysis that a formal transfer of data from the individual operator to a public body like EFSA is required. This generation and control of the pivotal data at the formative stages of risk regulation debilitates public involvement in the framing of risk during risk analysis; notwithstanding subsequent obligations to share the analysis with public authorities like EFSA, and with the general public to a limited extent.

It was found that the implementation of the precautionary principle in the current EU framework for risk regulation of GMOs is bereft of the fundamental rationale behind the principle. With regard to this, the book argued that the regulation of safety concerns about GMOs in the EU continues to be well within the classical risk paradigm. The precautionary principle is most relevant at the limits of expert rationality, in situations where scientific understanding is beset with systemic ambiguities and scientific ignorance. Bearing this in mind, the regulatory determination of the applicability of the principle through expert recommendation in EU GMO regulation is a problematic proposition. In this regard, the Commission has insisted that precaution is to be triggered by a technical finding of uncertainty, and that the principle should be confined to the stage of risk assessment. This leads to a situation where the trigger of the principle is controlled by techno-scientific spaces, through an expressedly non-precautionary assessment. Over and above this, the chapter noted that the bulk of the decisions to release GM crops in the EU is purely based on the assessment report of EFSA, given the deadlock in voting during committee procedures that constitute risk management. This has meant that the regulation of release of GMOs in the EU has been anything but precautionary, and there has been extremely limited participation of groups outside of techno-scientific spaces.

An appropriate normative perspective of the precautionary paradigm within which a policy framework can be developed, as suggested by WRR (the Dutch council for government policy), requires a proactive approach to uncertainty and vulnerability of people, society and the natural environment through public participation. Both the conceptual core of the regulatory tool of risk, and the normative emphasis on a proactive approach to incertitudes, require participation of publics that are wider than techno-scientific communities. This involves social inputs and political control that are wider than a pure reliance on techno-scientific reason, and where science is subsidiary to the debate about the reasons for developing the technology in question. This wider range of inputs includes consideration of concerns about social value, a comparative appraisal of viable alternative technologies, and a connection between the public purpose and the plausibility of dangers from the technology. The chapter on global rules regarding GMOs enquired whether possible institutional innovations that involve public participatory mechanisms in safety regimes within a precautionary paradigm are permissible within relevant global rules. Despite classical risk dominating the SPS Agreement, a realm that currently dominates other global rules through judicial construction, it was found that there is a slim possibility of WTO legality for Members to introduce these regulatory innovations.

On the whole, this book contended that it is important that law maintains efficacious channels to identify and respond to public values about regulation of GMOs. This includes public values about safety concerns as well, given the normative nature of questions in risk regulation, and the fundamental incertitudes in basic scientific knowledge underlying such regulation. With regard to the pursuit of public values, two significant routes that involved varied levels of public participation were identified in the EU regulation of GMOs. First is the strategy of consumer autonomy, effectuated through labelling and coexistence measures, where consumers can choose between different agricultural systems. This gives the consuming publics an opportunity to participate through consuming or avoiding consuming GM products. This strategy predicates on an assumption of ethical plurality that can be sustained through an ethically neutral framework, where the consuming publics can take the decision about appropriate consumption themselves. However, the strategy of consumer autonomy and sustaining ethical pluralism will be sufficiently participatory only if the disagreements about values pertain only to the consumption of technology, and not its development. The imagination of consumer autonomy is predicated on an ethically neutral framework through which divergent values can be furthered simultaneously by publics of different value disposition. However, controversies encountered during coexistence make the potential to create such an ethically neutral framework itself suspect. These factors include controversies regarding which stream of agriculture will bear the costs of segregation, the ethical implications of a strictly economic basis for coexistence measures and the technical (im)possibilities of segregation. Some groups cite these controversies to term coexistence measures as a back-door entry for fait accompli situations, where large-scale introductions ensure significant inadvertent presence of GM in conventional and organic products.

The second arena identified in the book where public values about regulation of GMOs are formally taken into account is deliberation in public bioethics committees. Culminating in advisory reports, the deliberation here is an important regulatory strategy to identify and reflect on public values about GMOs. The book emphasized the necessary connections of public bioethics to the public sphere as opposed to restricting it to methodological justifications of analytical philosophy. Chapter 6 underlined the need to include a public scrutiny not only of the ethics of consumption of GMOs but also the institutional ethics of development of technology through these committees. Further, a lack of ethical scrutiny of the instability of the underlying knowledge used in risk regulation, and regarding the normative assumptions during the framing of risk analysis, would be an inadequate form of deliberation. Significantly, advice from the EGE and the Danish Bioethics Council has brought focus to these issues. Such focus, by itself, does not correspond to a fundamental problem posed by the expert-dominated committees about the lack of lay participation in them. This problem relates to fundamental objective of use of bioethics advice; that is, use of ethics as a separate component in regulation of techno-science itself anticipates the inadequacies of a reliance on the values of techno-scientific communities to make their research and development compatible with a range of public values. However, identification of technology assessment, environment impact assessment and utility assessment by the EGE and Danish Council is an important opening for a more public engagement with public values about research, development and use of GMOs. During their scrutiny, these bioethics committees underlined the importance of public participation in the suggested impact assessment of GMOs. Whether and how these recommendations will be implemented in the future in the EU regulation of GMOs is speculative, and the nature of public participation in these implementations, currently indeterminable.

While the book has in itself explored the space for public participation in EU regulation of GMOs, this is quite different from reflecting on whether public participation is practically possible. How to structure participatory regulation in the EU depends on a host of factors, including political will, and is not germane to the scope of this book. However, it is important to recall invocations of an important hurdle to public partici-pation, viz., ill informed and disengaged lay publics. In this context, perhaps, it is also a valid question posed by many groups as to 'why is it that citizens and the public are always asked to choose among alterna-tives which others have designed and presented to them? Why is it that citizens are not asked to specify the world they want and the alternative which they desire?'[849] Whether lay publics are seen as citizens who are capable of finding their own meanings or seen as intellectually vacuous, would be key here. Various EU policy documents appear to answer this normative question in an affirmative and positive manner, while they commit themselves to public participation. If there is indeed such a policy commitment, then the description in this book points to a distinct disconnect between the commitment and implementation in the legal regime. The challenge, then, for policy-makers is how to implement them through justiciable principles and procedures. It would be erroneous to assume that the lay public influence in social choices for the development of technology is an attack on the scientist. Much to the contrary, it can pose added meaning to the technological enterprise through a responsible and democratic manner of development of technology: 'in recent focus groups and other public engagement exercises on nanotechnology, par-ticularly in Britain, members of the public voiced their experience of not having any agency, and were then joined by nano-scientists being involved but unable to make a difference either'.[850] It is important that a wider debate influences the development and regulation of technologies like GMOs, and the challenges and opportunities such participation in regulatory deliberation offers to democratizing law could yet be crucial for our collective futures.

---

[849] John Dixon, 'How can public participation become real?' [1975] January/ February *Public Administration Review* 69, 70.

[850] Arie Rip and Tsjalling Swierstra, 'Nano-ethics as NEST-ethics: Patterns of moral argument about new and emerging science and technology' (2007) 1 *Nanoethics* 3, fn 2.

Bibliography

Adams J., *Risk* (UCL Press 1995).

Adams M.D., 'The Precautionary Principle and the rhetoric behind it' (2002) 5 *Journal of Risk Research* 301.

Alemanno A., 'Judicial enforcement of the WTO Hormones ruling within the European Community: Toward EC liability for the non-implementation of WTO Dispute Settlement decisions?' (2004) 45 *Harvard International Law Journal* 541.

Alemanno A., 'The European Food Safety Authority at five' (2008) 1 *European Food and Feed Law Review* 1.

Alessandrini D., 'GMOs and the crisis of objectivity: Nature, science and the challenge of uncertainty' (2010) 19 *Social and Legal Studies* 3.

Alexander J., 'Critical reflections on "reflexive modernization"' (1996) 8 *Theory, Culture and Society* 43.

Amicus Brief submitted to the WTO Dispute Settlement Panel on *EC–Biotech* (Lawrence Busch and others, 30 April 2004).

Amicus Brief submitted to the WTO Dispute Settlement Panel on *EC–Biotech* (Jointly filed by the Center for International Environmental Law, Friends of the Earth–US, Defenders of Wildlife, Institute for Agriculture and Trade Policy and Organic Consumers Association, US, 1 June 2004) http://www.ciel.org/Publications/WTOBiotech_AmicusCuriaeBrief_June04.pdf.

Applegate J.S., 'The taming of the Precautionary Principle' (2002) 27 *William and Mary Environmental Law Review* 13.

Arendt H., *On Revolution* (Penguin 1963).

Arnstein S., 'A working model for public participation' (1975) January/February *Public Administration Review* 70.

Asen R., 'Seeking the "counter" in counterpublics' (2000) 10 *Communication Theory* 424.

Asimov Laws, http://www.asimovlaws.com/articles.

Atik J., 'The weakest link: Demonstrating the inconsistency of "appropriate levels of protection" in *Australia–Salmon*' (2004) 24 *Risk Analysis* 483.

Atwood M., *Oryx and Crake* (Bloomsbury 2003).

Atwood M., *Year of the Flood* (Bloomsbury 2009).

Baker T. and Simon J. (eds), *Embracing Risk: The Changing Culture of Insurance and Responsibility* (University of Chicago Press 2002).

Bakhtin M., *The Dialogic Imagination: Four Essays by M. M. Bakhtin* (Michael Holquist ed., Michael Holquist and Caryl Emerson trs, Texas University Press 1981).

Bauer M., 'Technophobia: A misleading conception of resistance to new technology' in Martin Bauer (ed.), *Resistance to Technology* (Cambridge University Press 1995).

Baxi U., *Human Rights in a Posthuman World: Critical Essays* (Oxford University Press 2007).

Beck U., *Risk Society: Towards a New Modernity* (Sage 1992).

Beck U., 'The reinvention of politics: Towards a theory of reflexive modernization' in Ulrich Beck, Anthony Giddens and Scott Lash (eds), *Reflexive Modernization: Politics, Tradition and Aesthetics in the Modern Social Order* (Polity 1994).

Beck U., *Ecological Politics in an Age of Risk* (Polity 1995).

Beck U., 'Environment, knowledge and indeterminacy: Beyond modernist ecology' in Scott Lash, Bronislaw Szerszynski and Brian Wynne (eds), *Risk, Environment and Modernity: Towards a New Ecology* (Sage 1996).

Beck U., 'Politics of Risk Society' in Jane Franklin (ed.), *The Politics of Risk Society* (Polity 1998).

Beck U., *World Risk Society* (Polity 1999).

Bernstein P., *Against the Gods: The Remarkable Story of Risk* (John Wiley and Sons 1996).

Betlem G. and Brans E., 'The future role of civil liability for environmental damage in the EU' (2002) 2 *Yearbook of European Environmental Law* 183.

Bignami F., 'The democratic deficit in the European Community rule-making: A call for notice and comment in comitology' (1999) 40 *Harvard International Law Journal* 451.

Bijker W., *Of Bicycles, Bakelite and Bulbs: Toward a Theory of Sociotechnical Change* (MIT Press 1995).

Bijker W. and Law J. (eds), *Shaping Technology/Building Society: Studies in Sociotechnical Change* (MIT Press 1992).

Bimber B., *The Politics of Expertise in Congress: The Rise and Fall of the Office of Technology Assessment* (State University of New York Press 1996).

Binimelis R., 'Coexistence of plants and coexistence of framers: Is an individual choice possible?' (2008) 21 *Journal of Agricultural and Environmental Ethics* 437.

Biotechnology and Biosciences Research Council, *Ethics, Morality and Crop Biotechnology* (prepared by Roger Straughan and Michael Reiss, UK Department of Trade and Industry 1996).

Black J., 'Constructing and contesting legitimacy and accountability in polycentric regulatory regimes' (2008) 2 *Regulation and Governance* 137.

Bodiguel L. and Cardwell M., 'Genetically Modified Organisms and the public: participation, preferences and protest' in Luc Bodiguel and Michael Cardwell (eds), *The Regulation of GMOs: Comparative Approaches* (Oxford University Press 2010).

Bogner A. and Menz W., 'How politics deals with expert dissent: The case of ethics councils' (2010) 35 *Science, Technology and Human Values* 888.

Bora A., 'Techno-scientific normativity and the "iron cage" of Law' (2010) 35 *Science, Technology and Human Values* 3.

Bridgers M., 'Genetically Modified Organisms and the Precautionary Principle: How the GMO dispute before the World Trade Organization could decide the fate of international GMO regulation' (2004) 22 *Temple Environmental Law and Technology Journal* 171.

Brownlee K., 'Civil disobedience' in Edward N. Zalta (ed.), *The Stanford Encyclopedia of Philosophy* (Spring 2010 ed.) http://plato.stanford.edu/entries/civil-disobedience.

Brownsword R., *Rights Regulation, and the Technological Revolution* (Oxford University Press 2008).

Burchell G., Gordon C and Miller, P.M. (eds), *The Foucault Effect: Studies in Governmentality, with two lectures by and an interview with Michel Foucault* (Harvester Wheatsheaf 1991).

Burgess A., *Cellular Phones, Public Fears and a Culture of Precaution* (Cambridge University Press 2004).

Burgiel S.W., 'The Cartagena Protocol on Biosafety: Taking the steps from negotiation to implementation' (2002) 11 *Review of European Community and International Environmental Law* 53.

Butti L., *Implementing the Precautionary Principle: Approaches from Nordic Countries and the EU* (Earthscan 2007).

Calhoun C., 'The public sphere in the field of power' (2010) 34 *Social Science History* 30.

Callahan D., 'Why America accepted bioethics' (1993) 23 *Hastings Centre Report* 8.

Callon M., 'The increasing involvement of concerned groups in R&D policies: What lessons for public powers' in Aldo Geuna, Ammon Salter and Edward Steinmueller (eds), *Science and Innovation* (Edward Elgar 2003).

Cardwell M., 'The release of genetically modified organisms into the environment: Public concerns and regulatory responses' (2002) 4 *Environmental Law Review* 156.

Cassese S., 'Global standards for national administrative procedure' (2006) 68 *Law and Contemporary Problems* 109.

Castel R., 'From dangerousness to risk' in Jens O. Zinn (ed.), *Social Theories of Risk and Uncertainty* (Blackwell 2008).

Centre for the Study of Environmental Change, *Public Attitudes towards Agricultural Biotechnologies in Europe: Final Report of Project with Five Partner Country Teams (Spain, Italy, Germany, France and the UK)* (EU DG-Research 2000).

Ceruzzi P., 'Moore's Law and technological determinism: Reflections on the history of technology' (2005) 46 *Technology and Culture* 584, 593.

Chambers S., 'Deliberative democratic theory' (2003) 6 *Annual Review of Political Science* 307.

Chan K., 'War of the papayas', *China Daily* (Hong Kong, 8 September 2011) http://www.chinadaily.com.cn/hkedition/201109/08/content_136 45581.htm.

Charnovitz S., 'The moral exception in trade policy' (1998) 38 *Virginia Journal of International Law* 689.

Charnovitz S., 'An analysis of Pascal Lamy's proposal on collective preferences' (2005) 8 *Journal of International Economic Law* 449.

Cheyne I., 'The precautionary principle in EC and WTO law: Searching for a Common Understanding' (2006) 8 *Environmental Law Review* 257.

Cheyne I., 'Life after the Biotech Products dispute' (2008) 10 *Environmental Law Review* 52.

Codex Alimentarius Commission Procedural Manual (16th edn, Codex Alimentarius Commission 2006).

Cranor C., 'The normative nature of risk assessment' (1997) 8 *Risk, Health, Safety and Environment* 123.

Craven M., 'Unity, diversity and fragmentation of international law' (2003) 14 *Finnish Yearbook of International Law* 3.

Dahl R.A., *Polyarchy: Participation and Opposition* (Yale University Press 1971).

Danish Council of Ethics, *Utility, Ethics and Belief in Connection with the Release of Genetically Modified Plants* (Danish Council of Ethics 2007).

Davies G., 'Morality clauses and decision making in situations of scientific uncertainty: The case of GMOs' (2007) 6 *World Trade Review* 249.

de Certeau M., *The Practice of Everyday Life* (University of California Press 1984).

de Chazournes L.B., 'New technologies, the Precautionary Principle and public participation' in Theresa Murphy (ed.), *New Technologies and Human Rights* (Oxford University Press 2009).

de Sadeleer N., 'Procedures for derogations from the principle of approximation of Laws under Article 95 EC' (2003) 40 *Common Market Law Review* 889.

de Sadeleer N., 'The Precautionary Principle in EC health and environmental law' (2006) 12 *European Law Journal* 139.

Dean M., 'Sociology after society' in David Owen (ed.), *Sociology after Post-modernism* (Sage 1997).

Dean M., *Governmentality: Power and Rule in Modern Society* (Sage 1999).

Dean M. and Hindess B., *Governing Australia: Studies in Contemporary Rationalities of Government* (Cambridge University Press 1998).

Defert D., '"Popular life" and insurance technology' in Jens O. Zinn (ed.), *Social Theories of Risk and Uncertainty* (Blackwell 2008).

DEFRA, *Risk Management Strategy* (April 2002).

Department of Trade and Industry, *GM Nation? The Findings of the Public Debate* (UK Department of Trade and Industry 2003).

Devlin P., *The Enforcement of Morals: Maccabaean Lecture in Jurisprudence of the British Academy* (Oxford University Press 1959).

Devos Y. and others, 'Ethics in the societal debate on genetically modified organisms: A (re)quest for *sense* and *sensibility*' (2008) 21 *Journal of Agricultural and Environmental Ethics* 29.

Dewey J., *The Public and its Problems* (Holt 1927).

Diebold N.F., 'The Morals and Order exceptions in WTO law: Balancing the toothless tiger and the undermining mole' (2007) 11 *Journal of International Economic Law* 43.

Dillon M., 'Sovereignty and governmentality: From the problematics of the "New World Order" to the ethical problematic of the World Order' (1995) 20 *Alternatives* 323.

Dixon J., 'How can public participation become real?' (1975) January/February *Public Administration Review* 69.

Dodds S. and Thomson C., 'Bioethics and democracy: Competing roles of National Bioethics organization' (2006) 20 *Bioethics* 326.

Doherty M., 'The application of Article 95 (4)–(6) of the EC Treaty: Is the Emperor still unclothed?' (2008) 8 *Yearbook of European Environmental Law* 48.

Dona A. and Arvanitoyannis I.S., 'Health risks of genetically modified foods' (2009) 49 *Critical Reviews in Food Science and Nutrition* 164.

Donzelot J., 'The power of political culture' (1979) 5 *Ideology and Consciousness* 71.

Douglas M., *Risk Acceptability According to the Social Sciences* (Routledge and Kegan Paul 1985).

Douglas M., 'Risk as a forensic resource' (1990) 119 *Daedalus* 1.

Douglas M., *Risk and Blame: Essays in Cultural Theory* (Routledge 1992).

Douglas M. and Wildavsky A., *Risk and Culture* (University of California Press 1982).

Douma W., 'Fleshing out the Precautionary Principle by the Court of First Instance – Analysis' (2003) 15 *Journal of Environmental Law* 372.

Doyle C., 'Gimme shelter: The necessary element of GATT Article XX in the context of the *China Audio-visual products* case' (2011) 29 *Boston University International Law Journal* 143.

Dratwa J., 'Taking risks with the Precautionary Principle: Food (and the environment) for thought at the European Commission' (2002) 4 *Journal of Environmental Policy and Planning* 197.

Dryzek J.S., *Deliberative Democracy and Beyond: Liberals, Critics, Contestations* (Oxford University Press 2002).

Dryzek J.S., Goodin R.E., Tucker A. and Reber B., 'Promethean elites encounter precautionary publics: The case of GM foods' (2009) 34 *Science, Technology and Human Values* 263.

Dryzek J.S. and Niemeyer S., *Foundations and Frontiers of Deliberative Governance* (Oxford University Press 2010).

Eastham K. and Sweet J., 'Genetically modified organisms: The significance of gene flow through pollen transfer' (Environmental issue report No. 28, European Environment Agency 2002).

Editorial, 'Do seed companies control GM crop research?' (2009) *Scientific American Magazine* (August, published 21 July), http://www.scientificamerican.com/article.cfm?id=do-seed-companies-control-gm-crop-research.

Editorial, 'Open to all' (2010) 465 *Nature* 10.

Eley G., 'Nations, publics, and political cultures: Placing Habermas in the nineteenth century' in Craig Calhoun (ed.), *Habermas and the Public Sphere* (MIT Press 1996).

Elliot A., 'Beck's sociology of risk: A critical assessment' (2002) 36 *Sociology* 293.

Ellis J., 'Overexploitation of a valuable resource? New literature on the Precautionary Principle' (2006) 17 *European Journal of International Law* 445.

Ellul J., *The Technological Society* (John Wilkinson tr, Knopf 1964).

Ellul J., *The Technological Bluff* (Geoffrey W. Bromiley tr, Eerdmans 1990).

Elster J. (ed.), *Deliberative Democracy* (Cambridge University Press 1998).

Engels J.M.M., Dempewolf H. and Henson-Appollonio V., 'Ethical considerations in agro-biodiversity research, collecting, and use' (2011) 24 *Journal of Agricultural Environmental Ethics* 107.

English M.R., 'Environmental risk and justice' in Timothy McDaniels and Mitchell J. Small (eds), *Risk Analysis and Society: An Interdisciplinary Characterization of the Field* (Cambridge University Press 2004).

Epps T., 'Reconciling public opinion and WTO rules under the SPS Agreement' (2008) 7 *World Trade Review* 359.

Epps T. and Green A.J., 'The WTO, science and the environment: Moving towards consistency' (2007) 10 *Journal of International Economic Law* 285.

Ericson R.V. and Doyle A. (eds), *Risk and Morality* (University of Toronto Press 2003).

Estlund D., 'Jeremy Waldron on law and disagreement' (2000) 99 *Philosophical Studies* 111.

Etty T., 'Current survey of substantive European Community law: Biotechnology' (2005) 5 *Yearbook of European Environmental Law* 293.

Etty T., 'Current survey of substantive European Community law: Biotechnology' (2006) 6 *Yearbook of European Environmental Law* 252.

Etty T., 'Current survey of substantive EC Environmental law: Biotechnology' (2007) 7 *Yearbook of European Environmental Law* 245.

EU Committee of Regions, 'Opinion of the Committee of the Regions on the Communication from the Commission to the Council and the European Parliament report on the implementation of national measures on the coexistence of Genetically Modified crops with conventional and organic farming' (2007) OJ C 57/11.

EurActiv, 'Cracks start to show in EU GMO policy' (6 April 2006), http://www.euractiv.com/cap/cracks-start-show-eu-gmo-policy-news-21 6526.

EurActiv, 'EU meeting on GM maize ends in deadlock' (26 February 2009), http://www.euractiv.com/sustainability/eu-meeting-gm-maize-ends-deadloc-news-221386.

EurActiv, 'EU considers "pause for thought" on GMOs' (15 April 2013), http://www.euractiv.com/en/biotech/eu-considers-pause-thought-gmos/article-168053.

EurActiv, 'GMO debate continues to divide EU' (no date), http://www.euractiv.com/climate-environment/gmo-debate-continues-divide-eu/article-170387Biotechnology/PIFB_StakeholderForum_Process.pdf.

European Commission, 'Commission Communication to Parliament and the Council Promoting the competitive environment for the industrial activities based on biotechnology within the community' SEC (91) 629 final.

European Commission, 'White Paper on Food Safety' COM (99) 719.

European Commission, *Attitudes of European Citizens Towards Environment* (Special Barometer 2952008).

European Commission, Commission archives http://cordis.europa.eu/foresight/research.htm.

European Commission, 'Communication from the Commission on the Precautionary Principle' COM (2000) 1 final.

European Commission, 'Proposal for a regulation of the European Parliament and of the Council laying down the general principles and requirements of food law, establishing the European Food Authority, and laying down procedures in matters of food safety' COM (2000) 716.

European Commission, 'White Paper on Environmental Liability' 9.2.2000 COM (2000) 66 final.

European Commission, 'White Paper on European Governance' COM (2001) 428 final.

European Commission, *European Science and Technology* (Eurobarometer 55.22001).

European Commission, *Scenario for Co-existence of Genetically Modified, Conventional and Organic Crops in European Agriculture: A Synthesis Report* (Joint Research Centre 2002).

European Commission, 'Communication of the European Commission concerning Article 95 (paragraphs 4, 5 and 6) of the Treaty of European Community' COM (2002) 760.

European Commission, 'Life Sciences: A Strategy for Europe' COM (2002) 27 final.

European Commission, 'European Legislative Framework for GMOs in Place' (Commission Press Release) IP/03/1056, 22 July 2003.

European Commission, 'European Legislative framework for GMOs in Place' (Commission Press Release) IP/03/1056, 22 July 2003.

European Commission, 'Guidelines for the Development of National Strategies and Best Practices to Ensure the Coexistence of Genetically Modified Crops with Conventional and Organic Farming' (Recommendation) [2003] OJ L 189/36.

European Commission, *European Science and Technology* (2nd edn, Eurobarometer 58.0 2003).

European Commission, 'European Action Plan for Organic Food and Farming' COM (2004) 415 final.

European Commission, 'Annex to the Report on the implementation of national measures on the Coexistence of GM crops with conventional and organic farming' (CommissionStaff Working Document) SEC (2006) 313.

European Commission, 'Proposal for a Council Decision concerning the provisional prohibition of the use and sale in Austria of genetically modified maize' COM (2006) 510 final.

European Commission, 'Report on the implementation of Member States' Coexistence Measures' (Communication) COM (2006) 104.

European Commission, 'Report on the implementation of Member States' Coexistence Measures' (Communication) COM (2006) 104.

European Commission, 'Report on the Implementation of National Measures on the Coexistence of Genetically Modified Crops with Conventional and Organic Farming' COM (2006) 104 final.

European Commission, 'Communication from the Commission to the European Parliament, the Council, the economic and social committee and the Committee of the regions on the freedom for Member States to decide on the cultivation of genetically modified crops' COM (2010) 375 final.

European Commission, 'Communication from the Commission to the European Parliament, the Council, the Economic and Social Committee and the Committee of the Regions on the Freedom for Member States to Decide on the Cultivation of Genetically Modified Crops' COM (2010) 380 final.

European Commission, 'Guidelines for the Development of National Coexistence measures to avoid the unintended presence of GMOs in Conventional and Organic crops' (Recommendation) [2010] OJ C 200/1.

European Commission, 'Proposal for a Regulation of the European Parliament and of the Council amending Directive 2001/18/EC as regards the possibility for the Member States to restrict or prohibit the cultivation of GMOs in their territory' COM (2010) 375 final.

European Council, Minutes of 2773rd Environmental Council Meeting of 18 December 2006, rejecting *European Commission Proposal for a Council Decision concerning the provisional prohibition of the use and sale in Austria of genetically modified maize*, COM (2006) 510 final.

EU Council, Press Release 6199/08 of 2849th meeting of the Council on Agriculture and Fisheries, 18 February 2008, http://www.consilium.europa.eu/ueDocs/ cms_Data/docs/pressData/en/agricult/ 98819.pdf.

European Food Safety Authority, 'Guidance on the environmental risk assessment of genetically modified plants' (EFSA 2010).

European Group on Ethics in Science and New Technologies, *EGE Proceedings of the Roundtable on Ethics of Modern Developments in*

Agricultural Technologies, 18 June 2008 (EGE 2008), http://ec. europa.eu/bepa/european-group-ethics/docs/publications/agriculture_ technologies_ethics.pdf.

European Group on Ethics in Science and New Technologies, *Opinion No. 24 on Ethics of Modern Developments in Agricultural Technologies* (EGE 2008) http://ec.europa.eu/bepa/european-group-ethics/docs/opinion 24_en.pdf.

European Group on Ethics in Science and New Technologies, *Opinion No. 25 on Ethics of Synthetic Biology* (EGE 2009) http://www. europarl.europa.eu/charter/pdf/text_en.pdf.

European Parliament, *Resolution on Genetically Modified Maize*, final edition, 8 April 1997.

European Parliament, *Resolution on Institutional and Legal Implications of the Use of 'Soft Law' Instruments*, 4 September 2007, 2007/ 2028(INI) P6TA (2007) 0366.

European Parliament, *Resolution on the Mandate of the Group of Advisers on the Ethical Implications of Biotechnology to the European Commission 1997* B4-0484/971997.

Evans J.H., 'Between technocracy and democratic legitimation: A proposed compromise position for common morality public ethics' (2006) 31 *Journal of Medicine and Philosophy* 213.

Ewald F., 'Insurance and risk' in Jens O. Zinn (ed.), *Social Theories of Risk and Uncertainty* (Blackwell 2008).

Expert Group on Science and Governance, *Taking European Knowledge Society Seriously: Report of the Expert Group on Science and Governance to the European Commission* (EUR 22700 2007).

Ezrahi Y., *Descent of Icarus: Science and the Transformation of Contemporary Society* (Harvard University Press 1990).

FAO/WHO, *Report of the 30th Session of the Joint FAO/WHO Codex Alimentarius Commission* Rome: Joint FAO/WHO Food Standards Programme ALINORM 07/30/REP, 2–7 July 2007.

FAO/WHO, *Report of the 38th Session of the Codex Committee on Food Hygiene* Joint FAO/WHO, Food Standards Programme ALINORM 07/30/13, 4–9 December 2006.

Feddersen C.T., 'Focusing on substantive law in international economic relations: The public morals of GATT's Article XX (a) and "conventional" rules of interpretation' (1998) 7 *Minnesota Journal of Global Trade* 75.

Fisher E., 'Evolution and status of the Precautionary Principle in international law' (2002) 4 *Environmental Law Review* 258.

Fisher E., 'Precaution, precaution everywhere: Developing a "Common Understanding" of the Precautionary Principle in the European Community' (2002) 9 *Maastricht Journal of European and Comparative Law* 7.

Fisher E., Jones J. and von Schomberg R., 'Implementing the precautionary principle: Perspectives and prospects' in Elizabeth Fisher, Judith S. Jones and Rene von Schomberg (eds), *Implementing the Precautionary Principle: Perspectives and Prospects* (Edward Elgar 2006).

Fisher E., Jones J. and von Schomberg R. (eds), *Implementing the Precautionary Principle: Perspectives and Prospects* (Edward Elgar 2006).

Flear M. and Vakulenko A., 'A human rights perspective on citizen participation in the EU's governance of new technologies' (2010) 10 *Human Rights Law Review* 661.

Fleurke F., 'What use for Article 95 (5) EC? An analysis of *Land Oberösterreich and Republic of Austria* v. *Commission*' (2008) 20 *Journal of Environmental Law* 267.

Follesdal A., 'The legitimacy challenges for new modes of governance: Trustworthy responsiveness' (2011) 46 *Government and Opposition* 81.

Follesdal A. and Hix S., 'Why there is a democratic deficit in the EU: A response to Majone and Moravcsik' (2006) 44 *Journal for Common Market Studies* 533.

Foster C.E., 'Public opinion and the interpretation of the WTO agreement on Sanitary and Phytosanitary measures' (2008) 11 *Journal of International Economic Law* 427.

Foucault M., 'Governmentality' in Graham Burchell, Colin Gordon and Peter M. Miller (eds), *The Foucault Effect: Studies in Governmentality, with two lectures by and an interview with Michel Foucault* (Harvester Wheatsheaf 1991).

Foucault M., 'What is Enlightenment?' in James Faubion (ed.), *Essential Works of Foucault 1954–84* (The New Press 1997).

Franklin J., 'Politics and risk' in Gabe Mythen and Sandra Walklate (eds), *Thinking Beyond Risk Society* (Open University Press 2006).

Fraser N., 'Politics, culture, and the public sphere: Toward a postmodern conception' in L.J. Nicholson and S. Seidman (eds), *Social Postmodernism: Beyond Identity Politics* (Cambridge University Press 1995).

Freestone D. and Hey E., *The Precautionary Principle and International Law: The Challenge of Implementation* (Kluwer 1996).

Friant-Perrot M., 'The EU regulatory regime for GMOs and its integration into Community food law and policy', in Luc Bodiguel and Michael Cardwell (eds), *The Regulation of GMOs: Comparative Approaches* (Oxford University Press 2010).

Fukuyama F., *Our Posthuman Future* (Picador 2002).

Fuller S., *The Governance of Science: Ideology and the Future of the Open Society* (Open University Press 2000).

Funtowicz S. and Ravetz J., 'Science for the post-normal age' (1993) 25 *Futures* 735.

Funtowicz S. and Ravetz J., 'Post-normal science: An insight now maturing' (1999) 31 *Futures* 641.

Furedi F., *Culture of Fear: Risk Taking and the Morality of Low Expectation* (Cassell 1997).

Gaines S.E., 'The WTO's reading of the GATT Article XX chapeau: A disguised restriction on environmental measures' (2001) 22 *University of Pennsylvania Journal of International Economic Law* 739.

Galligan D., 'Citizen's rights and biotechnology regulation' in Francesco Francioni (ed.), *Biotechnologies and International Human Rights* (Hart Publishing 2007).

Galloux J.C., Mortensen A.T., de Cheveigne S., Allansdottir A., Chatjouli A. and Sakellaris G., 'The institutions of bioethics' in Martin Bauer and G. Gaskell (eds), *Biotechnology: The Making of a Global Controversy* (Science Museum 2002).

Gardiner M.E., 'Wild publics and grotesque symposiums: Habermas and Bakhtin on dialogue, everyday life and public sphere' in Nick Crossley and John Michael Roberts (eds), *After Habermas: New Perspectives on the Public Sphere* (Blackwell 2004).

Gaskell G., Allum N., Stares S. and Allansdottir A., *Europeans and Biotechnology in 2005: Patterns and Trends – Final Report on Eurobarometer 64.3* (European Commission 2006).

Gee D. and Greenburg M., 'Asbestos: From magic to malevolent minerals' in Poul Harremoës (ed.), *The Precautionary Principle in the 20th century: Late Lessons from Early Warnings* (Earthscan 2002).

Georgiev V., 'Commission on the loose? Delegated lawmaking and comitology after Lisbon' (EUSA Twelfth Biennial International Conference, Boston, March 2011).

Gibbons M., *The New Production of Knowledge: The Dynamics of Science and Research in Contemporary Societies* (Sage 1994).

Glowka L., *Law and Modern Biotechnology: Selected Issues of Relevance to Food and Agriculture* (FAO legislative study 78, FAO 2003) ftp://ftp.fao.org/docrep/fao/006/y4839E/y4839E00.pdf.

GMO-free Europe, *GMO-freeRegions and Areas in Europe* (2009) http://www.gmofreeregions.org/fileadmin/files/gmofreeregions/full_list/List_GMOfree_regions_Europe_update_September_2010.pdf.

Goh G., 'GM food labelling and the WTO Agreements' (2004) 13 *Review of Community and International Environmental Law* 306.

Graham J.D., Johansson P.-O. and Nakanish J. 'The role of efficiency in risk management' in Timothy McDaniel and Mitchell J. Small (eds), *Risk Analysis and Society* (Cambridge University Press 2004).

Green D., 'From national health monopoly to national health guarantee' in David Gladstone (ed.), *How to Pay for Health Care: Public and Private Alternatives* (IEA Health and Welfare Unit 1997).

Gross M. and Krohn W., 'Society as experiment: Sociological foundations for a self-experimental society' (2005) 18 *History of the Human Sciences* 63.

Grossman M.R., 'Coexistence of genetically modified, conventional, and organic crops in the EU: The Community framework' in Luc Bodiguel and Michael Cardwell (eds), *The Regulation of GMOs: Comparative Approaches* (Oxford University Press 2010).

Group of Advisers on the Ethical Implications of Biotechnology, *Opinion 5 on Ethical aspects of the Labelling of Foods Derived from Modern Biotechnology* (EGE 1995) http://ec.europa.eu/bepa/european-group-ethics/docs/opinion5_en.pdf.

Grove-White R. and others, *Uncertain World: Genetically Modified Organisms and Public Attitudes in Britain* (Centre for the Study of Environmental Change 1997).

Guruswamy L., 'Sustainable agriculture: Do GMOs imperil biosafety?' (2002) 9 *Indiana Journal of Global Legal Studies* 461.

Guzman A.T., 'Global governance and the WTO' (2004) 45 *Harvard International Law Journal* 303.

Habermas J., *The Structural Transformation of the Public Sphere* (Polity 1989).

Habermas J., *Moral Consciousness and Communicative Action* (Polity 1992).

Hacking I., *The Emergence of Probability* (Cambridge University Press 1975).

Hacking I., *The Taming of Chance* (Cambridge University Press 1990).

Hagen P. and Weiner J., 'The Cartegena Protocol on biosafety: New roles for international trade in living modified organisms' (2000) 12 *Georgetown International Environmental Law Review* 697.

Hagendijk R., 'The public understanding of science and public participation in regulated worlds' (2004) 42 *Minerva* 41.

Hahn W., 'Technological assessment and the Congress' [1975] January/February *Public Administration Review* 76.

Halfon S., 'Confronting the WTO: Intervention strategies in GMO adjudication' (2010) 35 *Science, Technology and Human Values* 307.

Haraway D., *Simians, Cyborgs and Women: The Reinvention of Nature* (Free Association Books 1991).

Harding R. and Fisher E. (eds), *Perspectives on the Precautionary Principle* (Federation Press 1999).

Harremoës P. (ed.), *The Precautionary Principle in the 20th Century: Late Lessons from Early Warnings* (Earthscan 2002).

Harvey S., 'Regulation of GM products in a multi-level system of governance: Science or citizens' (2001) 10 *Review of European Community and International Environmental Law* 321.

Hay C. and Rosamond B., 'Globalisation, European integration, and the discursive construction of economic imperatives' (2002) 9 *Journal of European Public Policy* 147.

Hayles K., *How We Became Post Human: Virtual Bodies in Cybernetics, Literature, and Informatics* (Chicago University Press 1999).

Heiskanen V., 'The regulatory philosophy of international trade law' (2004) 38 *Journal of World Trade* 22.

Henderson H., 'Philosophical conflict: Reexamining the goals of knowledge' (1975) January/February *Public Administration Review* 77.

Herwig A., 'Whither science in WTO dispute settlement?' (2008) 21 *Leiden Journal of International Law* 823.

Hetman F., *Society and the Assessment of Technology* (OECD 1973) 85.

Hildebrandt M., 'A vision of ambient law' in Roger Brownsword and Karen Yeung (eds), *Regulating Technologies: Legal Futures, Regulatory Frames and Technological Fixes* (Hart Publishing 2008).

Hilgartner S., 'Thesocial construction of risk objects: or, how to pry open networks of risk' in James F. Short and Lee Clarke (eds), *Organizations, Uncertainties, and Risk* (Westview Press 1992).

Hirst P. and Thompson G., *Globalization in Question: The International Economy and the Possibilities of Governance* (Polity 1996).

Ho M., *Genetic Engineering or Nightmare? The Brave New World of Bad Science and Big Business* (Gateway Books 1998).

Hockman S., 'An international court for the environment' (2008) 11 *Environmental Law Review* 1.

Hoekman B. and Trachtman J., 'Continued suspense: *EC-Hormones* and WTO disciplines on discrimination and domestic regulation' (2010) 9 *World Trade Review* 151.

Hood C., 'Where extremes meet: "sprat" versus "shark" in public risk management' in Christopher Hood and David Jones (eds), *Accident and Design: Contemporary Debates in Risk Management* (UCL Press 1996).

Horst M. and Irwin A., 'Nations at ease with radical knowledge: On consensus, consensusing and false consensusness' (2010) 40 *Social Studies of Science* 105.

Howse R., 'China – measures affecting the protection and enforcement of intellectual property rights' (2011) 10 *World Trade Review* 87.

Howse R. and Horn H., 'European Communities – Measures affecting the approval and marketing of biotech products' (2009) 8 *World Trade Review* 49.

Howse R. and Regan D., 'The product/process distinction: An illusory basis for disciplining "unilateralism" in trade policy' (2000) 11 *European Journal of International Law* 249.

Howse R. and Tuerk E., 'The WTO impact on internal regulations – A case study of the Canada–EC asbestos dispute' in George Bermann and Petros Mavroidis (eds), *Trade and Human Health and Safety* (Columbia Studies in WTO Law and Policy, Cambridge University Press 2006).

Hudec R.E., 'Free trade, sovereignty, democracy: The future of World Trade Organization' (2002) 1 *World Trade Review* 211.

Hudec R.E., 'Science and post-discriminatory WTO law' (2003) 26 *Boston College of International and Comparative Law Review* 185.

Hunt J., 'The social construction of precaution' in Timothy O'Riordan and James Cameron (eds), *Interpreting the Precautionary Principle* (Earthscan 1994).

International Association for the Advancement of Artificial Intelligence *Interim Report from the Presidential Panel Chairs on Long-Term AI Futures* (IAAAI 2009) http:// research.microsoft.com/enus/um/people/horvitz/note_from_AAAI_panel_chairs.pdf.

International Law Commission, 58th Session, *Fragmentation of International Law: Difficulties Arising from the Diversification and Expansion of International Law*,Report of the Study Group of the International Law Commission, A/CN.4/L.682, 13 April 2006.

Irish Council of Bioethics, *Genetically Modified Crops and Food: Threat or Opportunity for Ireland? Opinion* (Irish Council of Bioethics 2005) 36 http://www.bioethics.ie/uploads/docs/GM%20Report1.pdf.

Irwin A., 'The politics of talk: Coming to terms with the New Scientific Governance' (2006) 63 *Social Studies of Science* 299.

Isaac G. and Kerr W., 'Genetically modified organisms and trade rules: Identifying important challenges for the WTO' (2003) 26 *World Economy* 28.

Jackson J.H., 'Dispute settlement and the WTO: Emerging problems' (1998) 1 *Journal of International Economic Law* 329.

James C., 'Global status of commercialized biotech/GM crops: 2006' (International Service for the Acquisition of Agri-biotech Acquisitions, ISAAA Brief # 35, 2006) http://www.isaaa.org/resources/publications/briefs/35/pptslides/Brief35slides.pdf.

Japp K.P. and Kusche I., 'Systems theory and risk' in Jens O. Zinn (ed.), *Social Theories of Risk and Uncertainty* (Blackwell 2008).

Jasanoff S., *Science at Bar: Law, Science and Technology in America* (Harvard University Press 1995).

Jasanoff S., 'Breaking the waves in science studies: Comment on H.M. Collins and Robert Evans, The Third Wave of Science Studies' (2003) 33 *Social Studies of Science* 389.

Jasanoff S., 'Biotechnology and empire: The global power of seeds and science' (2006) 21 *Osiris* 273.

Jasanoff S. and Wynne B., 'Science and decision-making' in Steve Rayner and Elizabeth L. Malone (eds), *Human Choice and Climate Change* (Battelle Press 1998).

Joerges C. and Neyer J., 'Politics, risk management, World Trade Organization governance and the limits of legalisation' (2003) 30 *Science and Public Policy* 219.

Joffe H., 'Risk: From perception to social representation' (2003) 42 *British Journal of Social Psychology* 55.

Jonas H., *The Imperative of Responsibility* (University of Chicago Press 1984).

Jones J., von Schomberg R. and Harding R., 'The precautionary principle and administrative constitutionalism: The development of frameworks for applying the precautionary principle' in Elizabeth Fisher, Judith Jones and Rene von Schomberg (eds), *Implementing the Precautionary Principle: Perspectives and Prospects* (Edward Elgar 2006).

Jonsen A.R., *The Birth of Bioethics* (Oxford University Press 1998).

Juma C., 'Thenew culture of innovation: Africa in the age of technological opportunities', Keynote Address (8th Summit of the African Union, Addis Ababa, 29 January 2007).

Kadvany J., 'Taming chance: Risk and the quantification of uncertainty' (1996) 29 *Policy Sciences* 1.

Kastenhofer K., 'Risk assessment of emerging technologies and post-normal science' (2011) 36 *Science, Technology and Human Values* 307.

Kemshall H., *Risk, Social Policy and Welfare* (Open University Press 2002).

Kemshall H., 'Social policy and risk' in Gabe Mythen and Sandra Walklate (eds), *Beyond the Risk Society* (Open University Press 2006).

Kitcher P., *Science, Truth and Democracy* (Oxford University Press 2001).

Kleinman D.L. (ed.), *Science, Technology and Democracy* (State University of New York Press 2000).

Kloppenburg R., *First the Seed: The Political Economy of Plant Biotechnology, 1491–2000* (University of Wisconsin Press 2004).

Knight F., *Risk, Uncertainty and Profit* (Hart, Schaffner and Marx essays, Houghton Mifflin Co. 1921).

Knorr-Cetina K., 'Transitions in post-social knowledge societies' in Eliezer Ben-Rafael and Yitzhak Sternberg (eds), *Identity, Culture and Globalisation* (Brill 2003).

Koller A., 'The public sphere and comparative historical research: An introduction' (2010) 34 *Social Science History* 261.

Koops C., 'Suspensions: To be continued, the consequences of AB Report in *Hormones II*' (2009) 36 *Legal Issues of Economic Integration* 353.

Korthals M., 'Ethical rooms for maneuver and their prospects *vis-à-vis* the current ethical food policies in Europe' (2008) 21 *Journal of Agricultural and Environmental Ethics* 249.

Kourany J.A., *Philosophy of Science after Feminism* (Oxford University Press 2010).

Krapohl S., 'Credible commitment in non-independent regulatory agencies: A comparative analysis of the European agencies for pharmaceuticals and food stuffs' (2004) 10 *European Law Journal* 518.

Krier J.E. and Gillete C.P., 'The un-easy case for technological optimism' (1985) 84 *Michigan Law Review* 405.

Krisch N., *Beyond Constitutionalism: The Pluralist Structure of Postnational Law* (Oxford University Press 2010).

Krisch N., 'Pluralism in post-national risk regulation: The dispute over GMOs and trade' (2010) 1 *Transnational Legal Theory* 1.

Krohn W. and Krucken G. (eds), *Riskante Technology ien: Reflexion and Regulation* (Suhrkap 1993).

Kuhn T., *The Structure of Scientific Revolutions* (University of Chicago Press 1962).

Kurzweil R., *The Singularity is Near* (Viking 2005).

Kysar D., 'Preferences for processes: The process/product distinction and the regulation of consumer choice' (2004) 118 *Harvard Law Review* 526.

Lamy P., 'The emergence of collective preferences in international trade: Implications for regulating globalisation' (15 September 2004) http://ec.europa.eu/archives/commission_1999_2004/lamy/speeches_articles/spla242_en.htm.

Lang A., 'Provisional measures under Article 5.7 of the WTO's agreement on Sanitary and Phytosanitary measures: Some criticisms of the jurisprudence so far' (2008) 42 *Journal of World Trade* 1085.

Lang A. and Scott J., 'The hidden world of WTO governance' (2009) 20 *European Journal of International Law* 575.

Lappe M. and Bailey B., *Against the Grain: The Genetic Transformation of Global Agriculture* (Earthscan 1999).

Latour, B., *We Have Never Been Modern* (Harvard University Press 1993).

Latour, B., *The Politics of Nature: How to Bring the Sciences into Democracy* (Harvard University Press 2004).

Lee M., 'Public participation, procedure, and democratic deficit in EC environmental law' (2004) 3 *Yearbook of European Environmental Law* 193.

Lee M., *EU Environmental Law: Challenges, Change and Decision-Making* (Hart Publishing 2005).

Lee M., *EU Regulation of GMOs: Law and Decision Making for a New Technology* (Edward Elgar 2008).

Lee M., 'The governance of coexistence between GMOs and other forms of agriculture: A purely economic issue?' (2008) 20 *Journal of Environmental Law* 198.

Lee M., 'Beyond safety? The broadening scope of risk regulation' (2009) 62 *Current Legal Problems* 242.

Lee M., 'Multi-level governance of GMOs in the European Union: Ambiguity and hierarchy' in Luc Bodiguel and Michael Cardwell (eds), *The Regulation of GMOs: Comparative Approaches* (Oxford University Press 2010).

Lee M., *EU Environmental Law, Governance and Decision Making* (Hart 2014).

Lee M., and Abbot C., 'The usual suspects? Public participation under the Aarhus Convention' (2003) 66 *Modern Law Review* 80.

Lee M., and Burrell R., 'Liability for the escape of GM seeds: Pursuing the "victim"?' 65 *Modern Law Review* 517.

Levidow L., 'Precautionary risk of Bt. Maize: What uncertainties?' (1997) 83 *Journal of Invertebrate Pathology* 113.

Levidow L. and Boschert K., 'Coexistence or contradiction? GM crops versus alternative agricultures in Europe' (2008) 39 *Geoforum* 174.

Levidow L. and Carr S., 'How biotechnology regulation sets a risk/ethics boundary' (1997) 14 *Agriculture and Human Values* 29.

Levitt M., 'Public consultation in bioethics: What's the point of asking the public when they have neither scientific nor ethical expertise?' (2003) 11 *Health Care Analysis* 15.

Livermore M.A., 'Authority and legitimacy in global governance: Deliberation, institutional differentiation, and the Codex Alimentarius' (2006) 81 *New York University Law Review* 766.

Lord C., 'Still in democratic deficit' (2008) 43 *Intereconomics: Review of European Economic Policy* 316.

Louet S., 'EC study reveals an informed public' (2001) 19 *Nature Biotechnology* 15.

Luhmann N., *Risk: A Sociological Theory* (A. de Gruyter 1993).

Luján J.L. and Todt O., 'Precaution in public: The social perception of the role of science and values in policy making' (2007) 16 *Public Understanding of Science* 97.

Lupton D., *Risk and Socio-cultural Theory* (Cambridge University Press 1999).

Lynn F.M., 'The interplay of science and values in assessing and regulating environmental risks' (1986) 11 *Science, Technology and Human Values* 40.

Maasen S. and Weingart P., 'What's new in scientific advice to politics' in Peter Weingart and Sabine Maasen (eds), *Democratization of Expertise? Exploring Novel Forms of Scientific Advice in Political Decision-Making* (Springer 2005).

Mackenzie R., Burhenne-Guilmin F., La Viña A. and Werksman J., *An Explanatory Guide to the Cartagena Protocol* (IUCN Environmental Policy and Law paper No. 46 2003).

Majone G. (ed.), *Regulating Europe* (Routledge 1996).

Majone G., 'Europe's democratic deficit' (1998) 4 *European Law Journal* 5.

Majone G., 'What price safety? The precautionary principle and its policy implications' (2002) 40 *Journal of Common Market Studies* 89.

Marceau G. and Trachtman J., 'The Technical Barriers to Trade Agreement, the Sanitary and Phytosanitary Measures Agreement, and the General Agreement on Tariffs and Trade' (2002) 36 *Journal of World Trade* 811.

Marchant G. and Mossman K., *Arbitrary and Capricious: The Precautionary Principle in the EU Courts* (AEI Press 2004).

Marwell J.C., 'Trade and morality: The WTO public morals exception after Gambling' (2006) 81 *New York University Law Review* 802.

Mayo D.G. and Hollander R.D. (eds), *Acceptable Evidence: Science and Values in Risk Assessment* (Oxford University Press 1991).

McAfee K., 'Beyond techno-science: Transgenic maize in the fight over Mexico's future' (2008) 39 *Geoforum* 148.

McGarity T. and Shapiro S., *Risk Regulation at Risk: Restoring a Pragmatic Approach* (Stanford University Press 2003).

McGrady B., 'Necessity exceptions in WTO law: Retreaded tyres, regulatory purpose and cumulative regulatory measures' (2009) 12 *Journal of International Economic Law* 153.

Meziani G. and Warwick H., *Seeds of Doubt* (Soil Association 2002).

Michalopoulos T., Korthals M. and Hogeveen H., 'Trading "ethical preferences" in the market: Outline of a politically liberal framework for the ethical characterization of food' (2008) 21 *Journal of Agricultural and Environmental Ethics* 3.

Minutes of meetings of the Environmental Council, Commission Press release IP/06/498, 27 June 2006, 12 April 2006; 2 December 2005 and 9 March 2006.

Misa T.J., 'Controversy and closure in technological change: Constructing "steel"' in Wiebe Bijker and John Law (eds), *Shaping Technology/
Building Society: Studies in Sociotechnical change* (MIT Press 1992).

Moore A., 'Public bioethics and public engagement: The politics of
proper talk' (2010) 19 *Public Understanding of Science* 197.

Moravcsik A., 'Despotism in Brussels? Misreading the European Union'
(2001) *Foreign Affairs* 114.

Moreno C., Todt O. and Lujan J.L., 'The context(s) of precaution:
Ideological and instrumental appeals to the precautionary principle'
(2010) 32 *Science Communication* 76.

Morris J., *Rethinking Risk and the Precautionary Principle*
(Butterworths-Heinemann 2000).

Moscovici S., 'The phenomenon of social representations' in Robert Farr
and Serge Moscovici (eds), *Social Representations* (Cambridge University Press 1984).

Mouffe C., *On the Political* (Routledge 2005).

Murphy J. and Levidow L., *Governing the Transatlantic Conflict over
Agricultural Biotechnology: Contending Coalitions, Trade Liberalisa-
tion and Standard Setting* (Routledge 2006).

Myhr A.I. and Traavik T., 'The precautionary principle: Scientific uncertainty and omitted research in the context of GMO use and release'
(2002) 15 *Journal of Environmental Ethics* 73.

Mythen G., *Ulrich Beck* (Pluto 2004).

Mythen G. and Walklate S., 'Introduction' in Gabe Mythen and Sandra
Walklate (eds), *Thinking Beyond Risk Society* (Open University Press
2006).

National Research Council (US), *Risk Assessment in the Federal Government: Managing the Process* (Committee on the Institutional Means for
Assessment of Risks to Public Health, National Academy Press 1983).

National Research Council (US), *Understanding Risk* (Committee on
Risk Characterization, Paul C. Stern and Harvey V. Fineberg eds,
National Academy Press 1996).

Neff R., 'The Cartagena Protocol and the WTO: Will the EU Biotech
products case leave room for the Protocol?' (2005) 16 *Fordham
Environmental Law Review* 261.

Nowotny H., Scott P. and Gibbons M., *Rethinking Science: Knowledge
and the Public in an Age of Uncertainty* (Polity Press 2001).

Nuffield Council for Bioethics, *Genetically Modified Crops: The Ethical
and Social Issues* (Nuffield Council 1999) http://www.nuffieldbio
ethics.org/sites/default/files/GM%20crops%20-%20full%20report.pdf.

Nzelibe J.O., 'Interest groups, power politics and the risks of WTO
mission creep' (2004)28 *Harvard Journal of Law and Public Policy* 89.

O'Malley P., *Risk, Uncertainty and Government* (Glasshouse 2004).

O'Malley P., 'Governmentality and risk' in Jens O. Zinn (ed.), *Social Theories of Risk and Uncertainty* (Blackwell 2008).

O'Neill O., 'Informed consent and genetic information' (2001) 32 *Studies in History and Philosophy of Biological and Biomedical Sciences* 689.

O'Riordan T. and Cameron J. (eds), *Interpreting the Precautionary Principle* (Earthscan 1994).

O'Riordan T. and Jordan A., *The Precautionary Principle, Science, Politics and Ethics* (Centre for Social and Economic Research on the Global Environment 1995).

O'Riordan T., Cameron J. and Jordan A. (eds), *Reinterpreting the Precautionary Principle* (Cameron May 2001).

Outhwaite W., *Habermas: A Critical Introduction* (Stanford University Press 1994).

Owens S., 'Engaging the public: Information and deliberation in environmental policy' (2000) 32 *Environment and Planning* 1141.

Pauwelyn J., 'The WTO Agreement on sanitary and phytosanitary (SPS) measures as applied in the first three SPS disputes EC – Hormones, Australia – Salmon and JapanApples' (1999) 2 *Journal of International Economic Law* 641.

Pauwelyn J., *Conflict of Norms in Public International Law* (Cambridge University Press 2003).

Peel J., 'Precaution – A matter of principle, approach or process?' (2004) 5 *Melbourne Journal of International Law* 1.

Peel J., Risk regulation under the WTO SPS Agreement: Science as an international normative yardstick?' (Jean Monnet Working Paper 02/2004) http://www.jeanmonnetprogram.org/papers/04.

Peel J., 'A GMO by any other name … might be an SPS risk! Implications of expanding the scope of the WTO Sanitary and Phytosanitary Measures Agreement' (2006) 17 *European Journal of International Law* 1009.

Peel J., 'When (scientific) rationality rules: (Mis)application of the precautionary principle in Australian mobile phone tower cases' (2007) 19 *Journal of Environmental Law* 103.

Peel J., 'Interpretation and application of the precautionary principle: Australia's contribution' (2009) 18 *Review of European Community and International Environmental Law* 1.

Peel J., *Science and Risk Regulation in International Law* (Cambridge University Press 2010).

Perrow C., *Normal Accidents: Living with High Risk Technologies* (Princeton University Press 1984).

Petersen A., Cath A., Hage M., Kunseler E. and van der Sluijs J., 'Post-normal science in practice at the Netherlands environmental assessment agency' (2011) 36 *Science, Technology and Human Values* 362.

Pew Charitable Trusts, 'The Stakeholder Forum on Agricultural Biotechnology: An Overview of the Process' (May 2003), http://www.pewtrusts. org/uploadedFiles/wwwpewtrustsorg/Reports/Food_and_Biotechnology/ PIFB_StakeholderForum_Process.pdf.

Picciotto S., 'Networks in international economic integration: Fragmented states and the dilemma of neo-liberalism' (1997) 17 *Northwestern Journal of International Law and Business* 1014.

Pieterman R., 'Culture in the risk society: An essay on the rise of a precautionary culture' (2001) 22 *Zeitschrift fur Rechtssoziologie* 145.

Poli S., 'The European Community and the adoption of international food standards within the Codex Alimentarius Commission' (2004) 10 *European Law Journal* 613.

Poli S., 'The EU risk management of GMOs and the Commission's defence strategy in the biotech dispute: Are they inconsistent?' in Francesco Francioni and Tulli Scovazzi (eds), *Biotechnology and International Law* (Hart 2006).

Ponti L., 'Transgenic crops and sustainable agriculture in the European context' (2005) 25 *Bulletin of Science Technology Society* 289.

Porter T., *Trust in Numbers: The Pursuit of Objectivity in Science and Public Life* (Princeton University Press 1995).

Prevost D., 'Opening Pandora's Box: The panel's findings in the EC-Biotech Products dispute' (2007) 34 *Legal Issues of Economic Integration* 67.

Qasim S.J.I., 'Collective action in the WTO: A "developing movement" toward free trade' (2008) 39 *University of Memphis Law Review* 153.

Ravetz J., 'Usable knowledge, usable ignorance: Incomplete science with policy implications' in William C. Clark and Robert Munn (eds), *Sustainable Development of the Biosphere* (Cambridge University Press 1986).

Reddy S., 'Claims to expert knowledge and the subversion of democracy: The triumph of risk over uncertainty' (1996) 25 *Economy and Society* 222.

Reid E. and Steele J., 'Free trade: What is it good for? Globalization, deregulation, and public opinion' (2009) 36 *Journal of Law and Society* 11.

Renn O., 'Concepts of risk: A classification' in Sheldon Krimsky and Dominic Golding (eds), *Social Theories of Risk* (Praeger 1992).

Renn O., 'Style of using scientific expertise: A comparative framework' (1995) 22 *Science and Public Policy* 147.

Rifkin J., *The Biotech Century* (Putnam 1998).

Rip A., 'The tension between fiction and precaution in nanotechnology' in Elizabeth Fisher, Judith S. Jones and Rene von Schomberg (eds), *Implementing the Precautionary Principle: Perspectives and Prospects* (Edward Elgar 2006).

Rip A. and Swierstra T., 'Nano-ethics as NEST-ethics: Patterns of moral argument about new and emerging science and technology' (2007) 1 *Nanoethics* 3.

Rissler J. and Mellon M., *The Ecological Risks of Engineered Crops* (MIT Press 1996).

Roberts J.M. and Crossley N., 'Introduction' in Nick Crossley and John Michael Roberts (eds), *After Habermas: New Perspectives on the Public Sphere* (Blackwell 2004).

Rodemeyer M., Sarewitz D. and Wilsdon J., *The Future of Technology Assessment* (Foresight and Governance Project, Woodrow Wilson International Center for Scholars December 2005) http://www.wilsoncenter.org/sites/default/files/techassessment.pdf.

Rodgers C., 'Implementing the community environmental liability directive: GMOs and the problem of unknown risk' in Luc Bodiguel and Michael Cardwell (eds), *The Regulation of GMOs: Comparative Approaches* (Oxford University Press 2010).

Roessler F., 'Appellate Body ruling in China-Publications and Audio-visual Products' (2011) 10 *World Trade Review* 119.

Royal Society (Great Britain), *Risk Assessment: Report of a Royal Society Study Group* (Royal Society 1983).

Safrin S., 'Treaties in collision? The Biosafety Protocol and the WTO Agreements' (2002) 96 *American Journal of International Law* 606.

Sagoff M., *The Economy of the Earth* (Cambridge University Press 1988).

Salter B. and Jones M., 'Biobanks and bioethics: The politics of legitimation' (2005) 12 *Journal of European Public Policy* 710.

Sandin P., 'Dimensions of the precautionary principle' (1999) 5 *Human and Ecological Risk Assessment* 889.

Sarewitz D., 'World view: Not by experts alone' (2010) 466 *Nature* 688.

Scholderer J., 'The GM foods debate in Europe: History, regulatory solutions, and consumer response research' (2005) 5 *Journal of Public Affairs* 263.

Sclove R., *Reinventing Technology Assessment: A 21st Century Model* (Science and Technology Innovation Program, Woodrow Wilson International Center for Scholars April 2010) http://www.wilsoncenter.org/sites/default/files/ReinventingTechnologyAssessment1.pdf.

Scotford E., 'Mapping the Art. 174 (2) EC case law: A first step to analyzing community environmental law principles' (2008) 8 *Yearbook of European Environmental Law* 1.

Scott J., 'On kith and kine (and crustaceans): Trade and environment in the EU and WTO' in Joseph Weiler (ed.), *The EU, the WTO, and the NAFTA: Towards a Common Law of International Trade?* (Oxford University Press 2000).

Scott J., 'European regulation of GMOs and the WTO' (2003) 9 *Columbia Journal of European Law* 213, 225–226.

Scott J., 'European regulation of GMOs: Thinking about judicial review in the WTO' (2004) 57 *Current Legal Problems* 140.

Scott J., 'The precautionary principle before the European Courts' in Richard Macrory, Ian Havercroft and Ray Purdy (eds), *Principles of European Environmental Law* (Europa Law Publishing 2004).

Scott J. and Sturm S., 'Courts as catalysts: Rethinking the judicial role in new governance' (2006) 13 *Columbia Journal of European Law* 565.

Scott J. and Vos E., 'The juridification of uncertainty: Observations on the ambivalence of the precautionary principle within the EU and the WTO' in Christian Joerges and Renaud Dehoussse (eds), *Good Governance in Europe's Integrated Market* (Oxford University Press 2002).

Sevenster H., 'The Environmental Guarantee after Amsterdam: Does the Emperor have new clothes?' (2000) 1 *Yearbook of European Environmental Law* 291.

Shaffer G. and Pollack M., 'Regulating between national fears and global disciplines: Agricultural biotechnology in the EU' (Jean Monnet Working Paper No.10 2004) http://centers.law.nyu.edu/jeanmonnet/papers/04/041001.pdf.

Shapin S. and Schaffer S., *Leviathan and the Air-Pump: Hobbes, Boyle and the Experimental Life* (Princeton University Press 1985, 2011 reissue).

Sheingate A.D., 'Promotion versus precaution: The evaluation of biotechnology policy in the United States' (2006) 36 *British Journal of Political Science* 243.

Shelley M., *Frankenstein; or, The Modern Prometheus* (Harding, Mavor and Jones 1818).

Siegrist M., 'The influence of trust and perceptions of risk and benefits on the acceptance of gene technology' (2000) 20 *Risk Analysis* 195.

Siipi H. and Uusitalo S., 'Consumer autonomy and availability of genetically modified food' (2011) 24 *Journal of Agricultural and Environmental Ethics* 147.

Silbergeld E., 'The uses and abuses of scientific uncertainty in risk assessment' (1986) 2 *Natural Resources and Environment* 17.

Silbey S. and Ewick S., 'Thearchitecture of authority: The place of law in the space of science' in Austin Sarat, Lawrence Douglas and Martha Umphrey (eds), *The Place of Law* (University of Michigan Press 2003).

Siltanen J. and Stanworth M., 'The politics of private woman and public man' in Janet Siltanen and Michelle Stanworth (eds), *Women and the Public Sphere: A Critique of Sociology and Politics* (Hutchinson 1984).

Slovic P. (ed.), *Risk Perception* (Earthscan 2000).

Sollie P., 'Ethics, technology development and uncertainty: An outline for any future ethics of technology' (2007) 5 *Journal of Information, Communication and Ethics in Society* 293.

Somsen H., 'Cloning Trojan horses: Precaution in reproductive technologies' in Karen Yeung and Roger Brownsword (eds), *Regulating Technologies* (Hart Publishing 2008).

Somsen H. and Etty T., 'ECJ: case report on Case C-236/01 *Monsanto Agricultura Italia Sp. A and Others* v. *Presidenza Del Consiglio dei Ministeri and Others*' (2004) 13 *European Environmental Law Review* 3.

Spinello R., *Cyberethics: Morality and Law in Cyberspace* (3rd edn, Jones and Bartlett 2006).

Starr C., 'Social benefit versus technological risk: What is our society willing to pay for safety?' (1969) 165 *Science* 1232.

Statutes of the Codex Alimentarius Commission, *Commission Procedural Manual 3* (14th edn, Codex Alimentarius Commission 2004), ftp://ftp.fao.org/codex/ProcManuals/Manual14e.pdf.

Steele J., 'Participation and deliberation in environmental law: Exploring a problem-solving approach' (2001) 21 *Oxford Journal of Legal Studies* 415.

Steele J., *Risks and Legal Theory* (Hart 2004).

Steele K., 'The precautionary principle: A new approach to public decision making?' (2006) 5 *Law, Probability and Risk* 19.

Steinberg R.H., 'The hidden world of WTO governance: A reply to Andrew Lang and Joanne Scott' (2009) 20 *European Journal of International Law* 1063.

Stirling A., 'Risk, uncertainty and precaution: Some instrumental implications from the social sciences' in Frans Berkhout, Melissa Leach and Ian Scoones (eds), *Negotiating Change* (Edward Elgar 2003).

Stirling A., 'Science, precaution and the politics of technological risk: Converging implications in evolutionary and social scientific perspectives' (2008) 1128 *Annals of the New York Academy of Sciences* 95.

Stirling A. and van Zwanenberg P., 'Risk regulation in Europe and the United States – A response to David Vogel' (2003) 3 *Yearbook of European Environmental Law* 47.

Stirling A., Renn O. and van Zwanenberg P., 'A framework for the precautionary governance of food safety: Integrating science and participation in the social appraisal of risk' in Elizabeth Fisher, Judith Jones and Rene von Schomberg (eds), *Implementing the Precautionary Principle: Perspectives and Prospects* (Edward Elgar 2006).

Strydom P., *Risk, Environment and Society: Ongoing Debates, Current Issues and Future Prospects* (Open University Press 2002).

Sunstein C., *Laws of Fear: Beyond the Precautionary Principle* (Cambridge University Press 2005).

Sykes A., 'Domestic regulation, sovereignty and scientific evidence requirements: A pessimistic view' (2004) 3 *Chicago Journal of International Law* 353.

Taylor C., 'Liberal politics and the public sphere' in Amitai Etzioni (ed.), *The New Communitarian Thinking* (University Press of Virginia 1995) 183.

Taylor-Gooby P. and Zinn J., 'Current directions in risk research, new developments in psychology and sociology' (2006) 26 *Risk Analysis* 397.

Teknologi-Radet, 'The Danish Board of Technology wins international award for public participation' (15 March 2011), http://www.tekno.dk/subpage.php3article=1735&survey=15&language=uk.

Tellichini M., 'Governing by values. EU ethics: Soft tools, hard effects'(2009) 47 *Minerva* 281.

Then C. and Potthof C., 'Risk reloaded: Risk analysis of genetically engineered plants within the EU – A report by Testbiotech e.V. Institute for Independent Impact Assessment in Biotechnology' http://www.testbiotech.org.

Thomison A., 'A new and controversial mandate for the SPS Agreement: The WTO Panel's interim report in the EC-Biotechdispute' (2007) 32 *Columbia Journal of Environmental Law* 287.

Tickner J.A. and Wright S., 'The Precautionary Principle and democratizing expertise' (2003) 30 *Science and Public Policy* 213.

Trouwborst A., *Evolution and Status of the Precautionary Principle in International Law* (Martinus Nijhoff 2002).

Trouwborst A., 'The precautionary principle in general international law: Combating the Babylonian confusion' (2007) 16 *Review of European Community and International Environmental Law* 185.

Tsioumani E., 'Genetically modified organisms in the EU: Public attitudes and regulatory developments' (2004) 3 *Review of European Community and International Environmental Law* 279.

Tulloch J., 'Culture and risk' in Jens O. Zinn (ed.), *Social Theories of Risk and Uncertainty* (Blackwell 2008).

Turnpenny J., Jones M. and Lorenzoni I., 'Where now for post-normal science? A critical review of its development, definitions, and uses' (2011) 36 *Science, Technology and Human Values* 287.

UK HMSO, *Report of the Committee of Inquiry into Human Fertilization and Embryology* (HMSO 1984) http://www.hfea.gov.uk/docs/Warnock_ Report_of_theCommittee_of_ Inquiry_into_Human_Fertilisation _and_ Embryology_1984.pdf.

UNEP, Report of the Open ended Ad-hoc Working Group of legal and technical experts on liability and redress in the context of the Cartagena Protocol on Biosafety on the Work of its Fourth Meeting UNEP/CBC/BS/WG-L&R/4/3, 13 November 2007.

US FDA Center for Food Safety and Applied Nutrition, *Report on Consumer Focus Groups on Biotechnology* (Alan S. Levy and Brenda M. Derby eds, Office of Scientific Analysis and Support 2000).

US House Committee on Science and Astronautics, 'Office of Technology Assessment background and status' (Committee print August 1973).

Valve H. and Kauppila J., 'Enacting closure in the environmental control of GMOs' (2008) 20 *Journal of Environmental Law* 339.

van den Bosche P., 'World Trade Organization dispute settlement in 1997 (Part I)' (1998) 1 *Journal of International Economic Law* 161.

van den Bosche P., 'World Trade Organization dispute settlement in 1997 (Part II)' (1998) 1 *Journal of International Economic Law* 479.

van der Zwaag D., 'The precautionary principle and marine environmental protection: Slippery shores, rough seas and rising normative tides' (2002) 33 *Ocean Development and International Law* 165.

Verheyen R., 'The Environmental Guarantee in practice: A critique' (2000) 9 *Review of European Community and International Environmental Law* 179.

Vogel D., 'Risk regulation in Europe and the United States' (2003) 3 *Yearbook of European Environmental Law* 1.

von Moltke K., 'Three reports on German environmental policy' (1991) 33 *Environment* 25.

von Schomberg R., 'The Precautionary Principle and its normative challenges' in Elizabeth Fisher, Judith Jones and Rene von Schomberg (eds), *Implementing the Precautionary Principle: Perspectives and Prospects* (Edward Elgar 2006).

Vos E., 'The role of comitology in European governance' in Deirdre Curtin and Ramses Wessel (eds), *Good Governance and the European Union – Concept, Implications and Applications* (Intersentia 2005).

Waldron J., *The Dignity of Legislation* (Cambridge University Press 1999).

Walker V., 'Keeping the WTO from becoming the World Trans-science Organization: Scientific uncertainty, science policy, and fact-finding in the Growth Hormones dispute' (1998) 31 *Cornell International Law Journal* 251.

Waltz E., 'Under wraps' (2009) 10 *Nature Biotechnology* 880, http://www.nature.com/nbt/journal/v27/n10/full/nbt1009-880.html.

Warner M., 'Publics and counterpublics' (2002) 88 *Quarterly Journal of Speech* 413.

Warnock M., *A Question of Life: The Warnock Report on Human Fertilization and Embryology* (Blackwell 1985).

Weiler J., 'The rule of lawyers and the ethos of diplomats: Reflections on the internal and external legitimacy of WTO dispute settlement' in Roger Porter, Pierre Sauve, Arvind Subramanian and Merico Beriglia Zampetti (eds), *Efficiency, Equity, Legitimacy: The Multilateral Trading System at the Millennium* (Brookings 2001).

Weingart P., *Anything Goes – rien ne va plus* (Kursbuch 1984).

Westrum R., *Technologies and Society: The Shaping of People and Things* (Wadsworth Publishing 1991).

WHO/FAO, *Understanding the Codex Alimentarius* (FAO 2005) www.fao.org/docrep/008/y7867e/y7867e/y7867e00.htm.

Wiener J.B. and Rogers M.D., 'Comparing precaution in the US and Europe' (2002) 5 *Journal of Risk Research* 317.

Winickoff D.E. and Bushey D.M., 'Science and power in global food regulation: The rise of the Codex Alimentarius' (2010) 35 *Science, Technology and Human Values* 356.

Winner L., 'Do artifacts have politics?' in Langdon Winner (ed.), *The Whale and the Reactor: A Search for Limits in an Age of High Technology* (University of Chicago Press 1986).

Winham G.R., 'International regime conflict in trade and environment: The Biosafety Protocol and the WTO' (2003) 2 *World Trade Review* 131.

World Commission on Environment and Development, *Our Common Future* (Oxford University Press 1987).

World Commission on the Ethics of Scientific Knowledge and Technology, *Expert Group Report on the Precautionary Principle* http://unesdoc.unesco.org/images/0013/ 001395/139578e.pdf.

WRR Scientific Council for Government Policy, *Uncertain Safety: Allocating Responsibilities for Safety* (Amsterdam University Press 2008).

Wu Mark, 'Free trade and the protection of public morals: An analysis of the newly emerging Public Morals Clause Doctrine' (2008) 33 *Yale Journal of International Law* 215.

Wynne B., *Risk Management and Hazardous Waste: Implementation and the Dialectics of Credibility* (Springer-Verlag 1987).

Wynne B., 'Frameworks of rationality in risk management towards the testing of naïve sociology' in Jennifer Brown (ed.), *Environmental Threats: Perception, Analysis and Management* (Belhaven Press 1989).

Wynne B., 'Risk and social learning: Reification to learning' in Sheldon Krimsky and Dominic Golding (eds), *Social Theories of Risk* (Westport 1992).

Wynne B., 'May the sheep safely graze? A reflexive view of the expert–lay knowledge divide' in Scott Lash, Bronislaw Szerszynski and Brian Wynne (eds), *Risk, Environment and Modernity: Towards a New Ecology* (Sage 1996).

Wynne B., 'Creating public alienation: Expert cultures of risk and ethics on GMOs' (2001) 10 *Science as Culture* 445.

Wynne B., 'Elephants in the room where publics encounter "science"? A response to Darrin Durant' (2008) 17 *Public Understanding of Science* 21.

Zander J., 'The "Green Guarantee" in the EC Treaty: Two recent cases' (2004) 16 *Journal of Environmental Law* 65.

Ziman J., 'Is science losing its objectivity?' (1996) 382 *Nature* 751.

Ziman J., *Real Science: What It Is, and What It Means* (Cambridge University Press 2000).

Zinn J.O., 'A comparison of sociological theorizing on risk and uncertainty' in Jens O. Zinn (ed.), *Social Theories of Risk and Uncertainty* (Blackwell 2008).

Zinn J.O., 'Risk Society and reflexive modernization' in Jens O. Zinn (ed.), *Social Theories of Risk and Uncertainty* (Blackwell 2008).

Zinn J.O., (ed.), *Social Theories of Risk and Uncertainty* (Blackwell 2008).

Zurek L., 'The European Communities Biotech dispute: How the WTO fails to consider cultural factors in the GM food debate' (2006) 42 *Texas International Law Journal* 345.

Index